To my most wonderful social media discovery ever …

Thank you for being my wife, my mentor, and my best friend, Peggy Lynn.

TWITTER FOR BUSINESS: TWITTER FOR FRIENDS

The Little Twitter Book You Should Not Tweet Without

For additional copies of Twitter for Business: Twitter for Friends, please contact:

Twitter for Business: Twitter for Friends
PO Box 4426
Topeka, KS 66604

Twitter: @tfbtff or @murnahan
Telephone: 1-866-293-2489 (U.S. and Canada)
Fax: 1-866-293-2489
Skype: murnahan
Website: twitterforbusinesstwitterforfriends.com

First Printing: July 2009

ISBN-13: 978-0-9824978-0-7

Table of Contents

Introduction:

I was asked to write this book *by* social media, and I wrote it *for* social media. My intentions for this book are not clouded by an aggressive hope for personal gain. Instead, I view my efforts the way social media works at its best, which dictates that if you see a great value in this information, you will use it and tell others. If the value is not there, my book sales will reflect this fact. This is a great wonder of social media, and it is based on the value to many people above the value to a few. If I do not prove any other benefit in my writing, it should be that providing value to others prevails in social media.

My background in social media marketing goes back to the mid-1990's and has been an amazing lesson in people, business, and the relationship between the two. I met many of my closest friends by way of social media, including my dearest friend and wife, Peggy, whom I met in 2000. Almost as a side-effect, I have also made the connections that have provided for earnings in the top fraction of a percent of money-earners. Yes, I called it a side-effect and you will soon understand why.

In the book you are about to read, I will discuss benefits of social media as well as things to avoid. I will share ways that you can enjoy Twitter more, and make it more manageable. Whether you seek to leverage your use of Twitter for your business, education, friendships, or other reasons, I believe you will find something useful here. I will also share why I credit Twitter as the most engaging social media tool ever created.

I present this book in a conversational style with information intended to help you, starting with using Twitter more efficiently, and then moving into advanced use of social media. Whether you are new to Twitter and setting up your first account, or you have used Twitter since its inception, this information will benefit you.

To date, I have made over 20,000 public updates on Twitter and I often reach the 1,000 direct message daily limit. In the process, I have made many fantastic business and personal connections.

I want to show you how to do the same, so let's get tweeting!

Twitter Now or Twitter Later

There is a lot of talk about social media marketing these days, and Twitter is in the spotlight. We have all heard it, and no industry is immune. All of the facts and figures point to the inevitability that your business will be affected by this turn in the marketing tide. This all got me to thinking about the dilemma that social media poses to each and every business entity.

The dilemma is in weighing the cost of participating versus the risk associated with not embracing social media marketing. So how can you mitigate your risk? Spend some time with me to understand why you should not overlook social media for another minute.

Social Media: The Big Marketing Shift

The part that may be the hardest for many companies is that this shift in the marketing tide has occurred during an already frightening time for business people. Companies who used to advertise in newspapers have found that they are largely ineffective in 2009. This has further added to the already obvious demise of print media. Similarly, television is losing the marketing battle at an astonishing rate. Your local affiliate stations used to receive a piece of the national advertisers spending, and that was cut by the major networks. This is all happening because the Internet has fully eclipsed all other media in both total advertising spend and consumer reach.

I can give you a long list of the things which have added to the social media boom, but I do not think I will need to explain this. You know it is here, and you are quite possibly feeling a bit overwhelmed by it all. It is a huge transition. Everybody is trying to find their way and hope when it all settles that they will have made the right choices. There is a fear for many of whether they will adapt to the new rules of marketing soon enough to be effective. This really is a dilemma of when to make a quantifiable effort: Will it be now, or will it be later? When should you adopt the new rules of engagement? Are you too late? It is too soon? The questions are so plentiful and pressing that I have watched it paralyze many would-be good business decision makers.

Social Media Dilemma: Risk vs. Risk

Perhaps you are just warming up to the idea that this "new" media is where things are going. The fear of jumping in is really pretty normal. After all, it is hard to believe in something after you have watched all the things you always knew about business and the economy suddenly change. Most of us were told that our home would always be a great investment. That seems to be a bit shaky now, although it will certainly return. We thought companies like General Motors, AIG, and others were unshakeable, and that the whole world economy could not all just collapse. Things have changed, and amidst all of that change around you, the thought of spending what seems like a fortune in order to effectively participate in a marketing method with a whole new set of measurement metrics probably feels a bit uncomfortable at best.

So what will make this all feel better and help it all make sense? The answer is this: If your competition does it sooner and better than you, the cost of lost opportunity will be greater than any other potential risk.

Social media marketing is truly not as new as you may see it on the surface. In many ways, it is the way it used to be done in every company, for as long as business has been done. The tools have changed a lot, but the communication basics are that if you develop a warm market, your business will always perform far better. Your sales process will be much smoother. Your brand image will be enhanced by the added customer satisfaction. The list of benefits to the added communication of social media marketing over advertising as usual should not be so hard to understand. All the same, as a social media and Internet marketing professional since the mid-1990s, I still often feel like I am trying to explain the color blue to a person devoid of sight.

A Picture of Social Media Marketing

I want to provide you with a mental picture of social media marketing. Work with me, please. Let's say that you are about to walk out the door to drive to your local Wal Mart for a couple of items. You are going to pick up a garden hose, some razor blades, a new alarm clock, and a few other items. As you head for the door with your car keys in hand, the doorbell rings. You get to the door and there is a salesperson standing there to greet you. They are with a company you have heard of, but you have never met this person. He says that he has all of those things you planned to pick up at Wal Mart. He has the garden hose, the razor blades, and even the alarm clock, and he happens to have them right there. He even has the brands you would buy. How does this feel to you? Are you a bit uneasy about it? Many people have answered this question for me, and it seems that the vast majority would still get in the car, drive to Wal Mart, sort through the aisles, wait in line, and return home. The trust factor compels them, and the guy at the door just did not have the trust yet, regardless of how hard he tried.

Setting a new stage from a social media perspective, I have a new picture for you. The person standing at the door is somebody you have had some brief communication with, and you realize you have some mutual friends. It warms up with a bit of friendly conversation, and what do you know? You belong to the same social group. It all starts to look different now, does it not?

The differences in these scenarios are very similar psychologically to an advertisement compared to a social media approach. In addition, with a social media approach, it is altogether likely that the phone rang before you even grabbed your car keys and a friend was on the other end to let you know they were sending the hose, razor, and alarm clock guy over. It has a completely different feel, and it is the reason that advertising has always been an uphill battle compared to proper relationship marketing.

Social Media Fears

Now that I have walked you through an analogy of digital social media compared to yesterday media, let us look at the worst social media dilemma of all. This time, you are in the selling position, and you are the guy at the door. Answer yourself this question: Do you want to be that guy at the door trying to peddle your goods, or would you rather be the hose, razor, and alarm clock guy who took the care to build relationships and will be walking up to the door already announced?

If you fear what happens if you embrace social media marketing today and that it may not work for you, the greater question should be in which hose, razor, and alarm clock salesperson you want to be. If you leave it up to your competition, your cost will be much greater indeed, because the deepest cost is that of missed opportunities.

The Twitter Basics

What is Twitter?

I suppose I should start out with just a short bit on what Twitter is, before I explain why you need it. Twitter is a communication tool by which users may post short messages to the Internet. Each Twitter user has the capability of aggregating other users' messages into one easy to review place that allows them to receive messages from whomever they choose. This could be news agencies, friends, industry leaders, or complete strangers they find interesting. By choosing to receive a users Twitter messages, referred to as "tweets", you select to "follow" their Twitter account. Once you follow their feed, they can see that you are interested, and conversely, you can see who follows your feed and shows interest in your messages.

The messages of users you follow may be viewed in aggregate on your Twitter home page or using one or more various tools developed to assist you in receiving and sending tweets. These tools include many developed for use on a Website, desktop, or mobile device.

When you use Twitter, you will find interesting people to follow, and presumably, many others will find you interesting as well, and choose to follow what you have to say. It is common practice for users to mutually follow others feeds, opening many opportunities to meet new and interesting people.

You do not need to know a lot about technology to use it, and you are likely to find that it is even easier to use than email. Once you get a few basics covered, it is not unlikely that you find it more useful than email, too! Because each tweet is limited in length to 140 characters, you can expect the communication to be concise.

If you are completely unfamiliar with Twitter, I invite you to see an example of the use of Twitter by reviewing my Twitter feed at the Web address as follows:

http://twitter.com/murnahan

Selecting a Username

You will be able to change your username later, so do not let this hold you back from the next steps. If you have already created your account, you can disregard the very first step. The first step is to consider your username. Some professionals will choose to use their first and last name in the form of first_last. This is not my favorite choice because for many people, including me, that is very long and harder to communicate verbally. Since you only have 140 characters to work with, it is best to keep the name very short. This becomes even more important when people retweet your messages, but we will get to that later.

As I mentioned, you can change your username later, however, I suggest giving careful thought to your username, as it will become your brand on Twitter. Some companies will elect to use a name recognized as their company, but it is more commonly accepted that your Twitter account has a real person behind it, even when it is used for business purposes. This is true with companies large and small, and is more profoundly observed with corporations such as Ford Motor Company. Within a corporation, there may be many Twitter users, and it is far friendlier for each to be uniquely identified as a person. After all, social media is largely successful because it shows a company's culture and its best assets, which are the people behind the brand.

Selecting a Profile Image

In choosing your profile image, also called an *"avatar"*, I recommend choosing an image you can stick with for an extended time. Similar to your username, the avatar will become an important part of your brand. When others who follow a lot of Twitter user's updates see your avatar, it is commonly how they will recognize you, even before your username. This should be something unique and easy to identify. Some things I frequently see are people wearing a red shirt, photos of dogs, close-up photos of an eye, and simple head and shoulder shots. Perhaps the one most teased are men wearing a business suit. In the social media world, these guys are often made fun of and have somewhat of a stigma about them. If you use an image of yourself, you should look clean and tidy, but just consider how much you want to look like an attorney. Even attorneys often do not like to look like attorneys. Be yourself and try to make it relaxed. As an example, my avatar is a casual snapshot taken during a relaxing weekend visit to Las Vegas.

During any service interruption on Twitter when avatars are unavailable, I have found it to require a lot more effort to carry on communications with others. The avatar could be likened to your face or your voice on Twitter. It is the most identifiable piece that others will use to communicate with you, so choose wisely. I also recommend that you not make frequent or significant changes to your avatar, for the same reasons stated here.

Most importantly, choose something to represent yourself other than the default avatar that is visible when your account is created. I have even seen others change the coloration of the default avatar, simply to prove that there is some level of engagement. If you take the lazy way and leave the default avatar in place, many people will not even give a second thought to looking further. Take these words seriously: Upload something the moment you set up your Twitter account. Even if you do not intend to keep it for more than a day or two, use something other than the default avatar. The default avatar is the greatest initial indication of accounts that were created using automated means and with the sole intention of sending worthless messages or to artificially inflate the follower count of some users who create them, or people who only use Twitter once.

Profile Design

The design of your Twitter home page can say a lot about you. In my case, it is not really exciting, but it matches very closely with the rest of my brand, including my blog, aWebGuy.com, and my corporation's Website, YourNew.com.

There are many great examples of unique background designs. A custom design is a way to stand out from those around you and to show a sense of who you are and what users may expect from getting to know you. Similar to your account avatar or biography, it should be uniquely yours and show a sense of individuality. Surf around a little and see what others have done. Seeking inspiration from experienced users may serve you well. If you have an established Website, I would consider it wise to use a design that complements your existing brand image.

Biography

Your first glimpse of my Twitter profile will tell you a bit about my personal brand. My present Twitter biography at time of writing this book reads as follows:

"I take coffee and cigarettes and turn them into better social media marketing and SEO. See me for a magical increase to your market share."

This is my brand, and it is personal to me. I have not yet met another person with a strikingly similar statement.

Your biography should be a statement that is unique to you. It should be interesting, or at least as interesting as you want others to expect from you. In my case, it conveys a sense of humor and also a sense of my professional life. This can be changed much easier than the avatar and the background, so take a chance, be creative, and have some fun with it. This will be an important piece that others will receive as their first impression, and you surely know the importance of a first impression. After they read it and decide whether to take the next steps to getting to know you, the biography is no longer such an important part of your brand identity. This is why I say that it is easier to change later. Biographies are limited to 160 characters, so choose those characters well.

Your First Tweets

In case you missed it, a "tweet" is a Twitter update. Each Twitter update is limited to 140 characters in length, so each tweet should be written concisely and with a direct meaning. I will discuss tweets more in another chapter, but before you go too far in seeking others to join and communicate with you using Twitter, consider creating a brief history. Like credit reporting, it is said that your history is a likely indication of future performance. With tweets, it is an important factor that others will use to see if you are interesting enough to hear more from and choose to follow your messages or not.

I suggest that you use your first tweets to lay groundwork of what others can expect. Without that foundation of a few tweets, others really only have an avatar, a background, and a blank slate upon which to base their opinions.

An interesting example of a first tweet was when Google joined Twitter. The first tweet by Google was in binary code. The tweet said "I'm 01100110 01100101 01100101 01101100 01101001 01101110 01100111 00100000 01101100 01110101 01100011 01101011 01111001 00001010", which translates to *"I'm feeling lucky"* in English. It provoked talk, and it was a good icebreaker for Google. You are not starting with a brand the size of Google, so I suggest starting with at least 20-25 tweets just to show a history and a bit of your personality. Even if you have no followers, you are not talking to yourself. You will be showing an indication of what others can expect from you.

Updates, Mentions, and Direct Messages

When you make a public update on Twitter, we call it a "tweet". You simply say what you like, and keep it under the 140 character limit. It can include a Website URL, a greeting, or whatever you like. There are also other types of messages, including a "mention", which contains a username preceded by the "@" symbol. Note that there is no space between the "@" and the username. An example of this would look similar to the message as follows:

@murnahan How are you today?

In this instance, the message would be in your public Twitter feed that anybody can see, but it would also deliver a message to me as a "mention". Twitter used to call this a "reply", but then changed the terminology to a "mention" because sometimes it is simply somebody referring to another person without actually replying directly to the user.

All public tweets are not only available in your public Twitter feed, but also available as individual pages. In order to view a tweet as an individual page, you may simply click on the timestamp just below a tweet on your Twitter home page. An interesting point on this matter is that they are relatively search engine friendly, and the body of the tweet will often be indexed by multiple search engines. This can be useful for brand recognition. However, I would add that as Twitter uses a "noindex" attribute in all links, it will not provide a benefit to algorithms such as Google PageRank™.

A direct message, commonly referred to as a "DM" is a message directed only to one intended recipient, and it is kept private between the sender and recipient. A direct message is sent in a form of the letter D then a space, and the username. An example of a direct message is as follows:

D murnahan How are you today?

An important note about direct messages is that you may only send them to recipients that follow your feed. I suggest being mindful of this fact when you send a direct message. If you send a message to somebody, you should be sure that you are following their feed in return. Even though you have sent them a direct message, they cannot reply to you with a direct message unless you are also following them.

Automated Direct Messages

There are multiple services which will allow you to send an automated direct message to each new follower of your Twitter feed. The automated direct message has so commonly been used to say "go see my Website" or advertising "how to gain more followers" that the Twitter community has sounded a revolt. I do not want to spend a lot of my time on this topic, so I will simply say: DON'T DO IT!

Automated direct messages have been the subject of many very angry debates, and are simply useless, in my opinion. If you want to greet every new follower, do it the right way. Send them a nice customized message welcoming their communication if you like.

Twitter Help

I hope this book will provide adequate help for many of your questions. However, if technology was always perfect, many people would be out of work. When something breaks or you have a technical issue come up, you should know where to turn.

In order to get the best of Twitter, I suggest becoming familiar with the Twitter Website (http://twitter.com). This is especially true if you will be placing a lot of confidence in the service by using it for business purposes.

Twitter has a help Website at http://help.twitter.com but in the spirit of Web2.0 camaraderie, you will commonly receive your best help by asking other users. If you pose a question about Twitter to other Twitter users, you can generally expect a fast answer. Of course, if you do not find it in this book, you are also welcome to send a message to @murnahan for help.

Anatomy of a Tweet: Avoiding Miscommunication

Although 140 characters may seem small, there is plenty of room for miscommunication. Obvious possibilities for miscommunication exist in the inherent lack of vocal intonation, facial expression, body movement, and language. We cannot avoid all possibilities of miscommunication, but I will offer you some points to remember.

If a tweet has a link to a Website address, I would recommend not commenting on the tweet unless you have actually visited the Website. It is common for people to use a headline from a Website when sending a tweet, and that headline may or may not accurately describe the sender's point of view. If it is a link to a blog, you may even find that the user has made a comment on the blog that states their viewpoint.

You should also be mindful that people will often retweet somebody else's tweet by adding "RT" followed by the originating username, or adding "via" followed by the originating username. Attributing content to the originating sender is a part of Twitter's rules. Since a retweet becomes a combination of multiple users input, it is often a good practice to click on the original username to see if the tweet has been edited and to see if the user you received the tweet from has made comments or edited the tweet. If it is an older tweet and you did not find it in their Twitter timeline, it may be a good time to become familiar with the Twitter advanced search functions to find the original tweet. I will expand on the Twitter retweet more in another section of this book.

Each tweet has a timestamp showing when it was sent. The timestamp is also a link to the individual tweet on a separate page. When you need more room to comment on a tweet, you can simply type your tweet and include the link to the other tweet. Although Twitter has a 140 character limit per tweet, this does not mean that you must say all that you have to say in a single tweet. It is much better to link to another tweet than to try and abbreviate to the point your message is mistaken.

Twitter is designed to provide for conversations. When a tweet is a reply to another tweet, it will state *"in reply to"* beside the timestamp. This text is a link, which will allow you to track the tweets back to view a complete conversation. If you join a conversation, it is best to take a moment to review what you have missed before you form an opinion. You may also find that people will sometimes send a tweet as a reply, simply because the reply link was easier to use than to type the username. This means that what may seem to be a reply to one topic may be about an entirely different topic. Conversely, you may find that something that was intended to be a reply may not have been sent as a reply. Being aware of this can help you to avoid miscommunications.

In the area of miscommunication, it is often best to use good judgment and communicate with other users one-on-one before making assumptions. This is a great time to use direct messages. If the user is not following your feed, it may be best to politely ask that they follow you so that you can communicate with them through direct message.

Hashtags

Hashtags are used to associate messages of the same topic and are a term prefixed by a "#" mark placed anywhere within an update. An example of a hashtag is the one I will associate with this book. Users referencing "Twitter for Business: Twitter for Friends" may use a hashtag that looks like this: #tfbtff. In many Twitter tools, hashtags create a hyperlink directly to a Twitter search for the term used. The hashtag may be useful for coordinating an event or topical discussion in which you wish for others to join in and find relevant information on that topic.

Although Hashtags are not registered, I believe that it is a good idea to review current hashtags being used before you commit to using a new hashtag. It could be confusing for others if you use a hashtag that is already being used for another topic. You may do this by performing a search for the hashtag in the advanced search function at http://search.twitter.com/advanced or at http://hashtags.org where you will find a directory of commonly used hashtags. When the hashtag you use is unique, it can be much simpler to filter information and track your topic.

FollowFriday and Other Memes

If you have not seen the term #FollowFriday in a message, it is only a matter of time. The #FollowFriday hashtag is an example of use of a hashtag to track recommendations by one user for another. It all started with the idea of recommending friends or people who were respected in the Twitter community, but grew into something that has been the topic of many complaints. The complaints center on excessive and reckless use of the hashtag for recommending a long list of people rather than a merit-based recommendation. I have seen this meme become the only use of Twitter by some users for much of the day on Fridays.

If you want to give somebody a proper FollowFriday recommendation, I suggest making it useful. Include why you think they are a good person to follow. Say something clever or meaningful and make it original. If you see a FollowFriday endorsement from me, it may just say something like "Follow Yoon (@yoonhoum) because he is sharp as a pitchfork".

It is best if you do not include more than one or two, but maximum of three people in your endorsement. Sending a message full of usernames will not do your friends any favors, and it will not make you more popular with most people.

I welcome FollowFriday endorsements, and I consider a real endorsement an honor. On the other hand, I consider it a disrespect of the purpose of Twitter as a communication tool when I see a huge volume of recommendations for people they never even met.

For more on the topic of #FollowFriday, I welcome you to read my blog on this topic at the URL as follows: http://tfbtff.com/KUl14

I have used the #FollowFriday meme as an example, but within the Twitter community, you will find many memes, or trends that catch on and spread quickly. Although Twitter has implemented a

"Trending Topics" area on the right side of the home page, it only displays the top ten trends of the moment. This information is also available on Twitter's search page at http://search.twitter.com. In order to see additional trending topics, I offer you these alternatives:

Retweet Radar

Retweet Radar presents the top current trends being retweeted (repeated) on Twitter and also shows a historical view with their "trend archives". Visit http://retweetradar.com for a closer look.

Twemes

Twemes has a "tag cloud" including recent trends based on words that are presently being used the most at Twitter. Visit http://twemes.com for a closer look.

Twitter Security: Use Caution, Not Paranoia

As with any tool, there is potential for misuse. Some people will use a kitchen knife to tighten a screw, while others will use a communication tool to loosen your security. You can minimize your exposure to outside threats by practicing diligence.

Password Creation

In my years of managing a wholesale Internet access company, I have seen enough bad passwords to fill a lifetime. To put the importance of password strength into perspective, I will tell you how easy your password may be cracked. This may sound a little geeky, but try to follow along.

In my business, we have often encountered clients who purchased the user accounts of another Internet service provider. This is normally not done with a small number of accounts, but rather a bulk purchase of thousands of user accounts. The purchase generally includes the name, address, email, payment information, and a username and password combination, but not always. Because of multiple types of authentication systems (e.g. PAP and CHAP), sometimes those passwords are not available in a decrypted format, but instead come to the purchaser as encrypted in a format that they cannot support. Thus, the purchaser of the accounts must either decrypt thousands of passwords, or alternately provide a means for all of the users to prove their identity and then create a new password. Since I am referring to the password for gaining access to the Internet, it is not as simple as a lost password request or emailing them a new password. After all, if they cannot connect to the Internet, how will they retrieve their password? In cases when the purchaser finds that there is not enough data to properly authenticate each user's identity, they must decrypt as many of the passwords as possible and then provide an alternate solution for the remaining users. In order to expedite this process, it has been our practice to step in and perform this process for the client. After all, the sooner we can move the users to our systems, the sooner we can begin selling our service.

This may sound terribly difficult for some people, and I warned that it may sound geeky. The point I am getting to is that my team has often cracked the passwords of thousands of users, and many within just minutes. Some projects have taken many hours, days, or even weeks, but we have been able to decrypt thousands of passwords per hour by comparing the encrypted version against a comprehensive dictionary file. Now if we can do that, the presumption that a hacker cannot do the same is a dangerous notion.

The more difficult you make your password; the less likely you are to become a victim. Sure, we are just talking about Twitter, but if you can prevent somebody else using your identity, it should be worth just a moment to understand and implement how to do so. Here are a few important tips:

- A password should never be a word, name, place, date, or phone number.

- Include letters and numbers.

- Include both lowercase and uppercase letters.

- Include special characters in your password, such as @, $, %, &, #, !, *, and others.

- Passwords that are easy to remember are easy to crack.

- Change your password regularly!

It is important to remember why we use passwords. We use passwords so that other people do not have access to our information, and our identity. If you make it harder for others to access your account, you decrease the odds that they will. I hope that you will take this into consideration with all of your accounts, and not only with Twitter. As a professional who engages in Internet security matters every day, I consider password security the single simplest yet overlooked risk of all.

Application Security

Along with the creation of the many applications created for use with Twitter comes a greater opportunity for fraud. The majority of the Websites requiring your Twitter login are doing so in order to provide a legitimate service, but it is often difficult to know the difference. There is not only a consideration of the application developer's intent, but also that of others who may have access to their systems, or yours. The most effective way to minimize your risk is to minimize your exposure. This means establishing a small group of providers that you trust, and limiting the number of services where you provide your account credentials. Rather than to make assumptions about the security of each of the many services relating to Twitter, I would suggest using caution and asking others' opinion. If a service is referenced in major social media or technology blogs such as Mashable.com or TechCrunch.com, there is a greater presumption of legitimacy. However, for every rule, there is an exception.

Watch What You Tweet

In most areas of my life, I tend to take caution to extremes. Since I am a cautious technology fanatic, I have three completely redundant monitored home security systems. In fact, I think I am the only guy on my block with a video surveillance system that simultaneously sends video and still photos to servers in three separate datacenters over cable, DSL, and wireless Internet services. Stopping short of explaining my level of caution, I will just say that my home is more secure than many banks, and has a very protective and heavily armed guard, me. This is not just out of paranoia, but it does relate to my career in the Internet services industry. I have operated Internet services for a long time, and there is always that outside chance that somebody is silly enough to think I have servers with sensitive data at my home.

Since I do not expect that most people will choose to go to the extremes of security protocols which I have adopted, a bit of sensible prevention is another alternative. Your level of personal security is a unique choice that you will make. I simply want to remind you that on the Internet, you can meet all types. Some are fortunate to meet their spouse online, like I did, and others meet some real creepy folks, which I have also done. In fact, I will never forget the Web hosting client who truly believed he was a vampire and made threats to come to my home to devour me.

With regard to Twitter, I take a very relaxed view of what I say. After all, I believe in being myself without apology. This will not be appropriate for everybody, and you should know that even if you delete an update, it may still be found in Twitter searches, other Websites that archive Tweets, and even including a Google cache. If you slip up and send something, but later change your mind, it may be too late. The best advice for this is to think before you tweet, and remember that it is there for the world to see.

Protected Tweets

If you have a reason that you only want a set group to receive your Twitter updates, you should be aware that there is an option to protect your updates. This will not often be productive for anybody hoping to meet a lot of new people or use the service for most business purposes. However, if you have reason to communicate with only a select group, this option may be set by clicking "Settings" and then checking the "Protect my updates" checkbox at the bottom of the page. Once you save the setting, only the people you approve to receive your updates will see your updates from that point forward.

Twitter Limits

Twitter implements a few limitations, largely for the purpose of curbing abuse of their systems. Most people would consider the limits to be generous, but others will become frustrated once those limits are reached. The Twitter usage limits, at the time this is written, are as follows:

Public Message Limits

There is an hourly public message limit, including standard updates and mentions, of 100 messages per hour. There is an additional cap of 1,000 public messages per day.

Direct Message Limits

The direct message, or "DM" limit is based on a daily cap of 1,000 messages. If you are very social, and if you are like me and respond to everybody who messages you, this is the limit that will get in the way the most. I have reached this limit many times.

API Limits

The API, or application program interface, is something behind the scenes, and is the method by which applications outside of Twitter's own systems communicate data with Twitter's systems. Although the API limit is the one I have most commonly heard complaints about, it is also an easy limit to address and manage. The API limit is capped at 150 requests per user to Twitter per hour across all third-party applications, including desktop applications, Web applications, and etcetera. This means that if you access your account using an outside program or multiple programs, you may only make 150 requests for updated information per hour.

I use two popular applications called TweetDeck and Seesmic Desktop. Because of my heavy use of Twitter, I have found that the default API settings would often cause me to reach the API limit. Fortunately, each of these applications allow for modifying the time between automated requests sent to Twitter for updates. Many applications will allow you to modify the update interval, making it very easy to manage your use of the API. I like to set the update intervals longer and manually refresh the data when I am ready. I find that Seesmic Desktop handles this well by allowing users to refresh data based on a specific category, such as private messages, mentions, public messages, and etcetera.

Twitter Tools

Along with the 2009 explosion of Twitter use, there have been many tools developed for use alongside Twitter. They have sprouted up all over, and it can be challenging to know which will hold a high value. Some of the most valuable tools I have found to use in conjunction with Twitter are as follows:

Twitter Advanced Search

Twitter advanced search (http://search.twitter.com/advanced) features allow you to target specific information with great accuracy. It is worth taking a little time to become familiar with the advanced search features.

The advanced search form allows you to enter many specific search criteria, or you may also manually use the search operators as follows:

Search for this:	Returns results for:
this that	both "this" and "that"
"this that"	exact phrase "this that"
this OR that	either "this" or "that"
#this	the hashtag "this"
from:username	tweets sent by "username"
to:username	tweets sent to "username
@username	tweets mentioning "@username"

near:London	tweets occurring near London
near:Paris within:10mi	tweets within 10 miles of Paris
this since:2009-07-25	"this" since July 25th
this until:2009-07-25	"this" prior to July 25th
this -that	"this" but not "that"
this ?	"this" and a question
this filter:links	"this" containing a hyperlink
this source:seesmic	"this" sent through Seesmic

If you try some of these search operators together, you may be surprised what you can find using Twitter. It is often much easier to simply use the advanced search page, but you may also use these functions in your desktop client such as Seesmic Desktop or TweetDeck, which I will explain in another section. This may be a good time to jot a note in the back of the book as a reminder of specific searches that could help you stay informed about topics important to you.

URL Shortening

A URL (Web address) shortening service is an invaluable tool for fitting a long Web address into the short 140 character limit of Twitter. Most Twitter applications have this tool built in, but there is a better way to perform the task of shortening URL.

Bit.ly (http://bit.ly) is a service that will free up a lot of space to say more in your message, and will also add great tracking benefits to measure usage of the shortened URL. I choose bit.ly for a few very specific reasons. One of my top reasons is that it provides a permanent "301" redirect of the shortened URL. A 301 redirect is a type of URL redirection that provides a far more search engine friendly result. Additionally, bit.ly provides tracking of the URL including a real-time click through count, reports of the past day, week, month, and total traffic. When you use bit.ly to shorten a URL, it will track your uniquely created URL and also an aggregate traffic report for all others who shortened the URL, and also their username. This data is very useful to measure whether your message is being heard.

Follower Management

When you want to receive updates from other Twitter users, you "follow" them, and if they want to receive updates from you, they choose to "follow" you in return. If you become a highly active Twitter user, you may find challenges in managing those you follow and who follow you. Many will elect to use a third party tool to manage the task.

What I will tell you is likely not what many advanced users may expect me to say, but the Twitter Website is the most reliable tool I have found for managing people following your feed and whom you follow in return. I have tried some of the other tools for managing reciprocal following and for bulk following others. I have found that each of the services has made errors on more than one occasion.

The exception I make for the opinion above is the use of Tweepular (http://tweepular.com). I use Tweepular to create a list of users who follow me but that I am not yet following. I use this list to review users whom I may wish to get to know.

Tweepular also allows users to sort followers based on criteria such as finding users that are not following in return. Since so many people will follow you only to build their follower base, and then drop you as soon as you re-follow, Tweepular can help you to discover those users and unfollow them. I watch for this trend, and if I see them follow me again at a later time, I will often block them, because their intent has become obvious to me. This is a common practice of people seeking to inflate their follower count and presumably increase their credibility. This is a practice that is frowned upon and provides little real value. I discourage this practice.

I have found that it is far more effective to build strong relationships with a smaller group than to be concerned with fast growth. I will discuss this topic in much greater length later.

Who's Online and When?

Twitter Analyzer (http://twitteranalyzer.com) can answer questions of when people are online, whom you communicate with the most, if there is somebody you may be neglecting, and much more. I believe that the best way for you to become familiar with this excellent tool is to actually start using it. Be sure to review each link there, and you will begin to understand why I find it extremely useful.

What's Happening in Your Area

Whether you are in law enforcement and trying to find witnesses to an auto accident, or hungry and looking for ideas about where to go for dinner, Nearby Tweets is a handy tool. It is a simple way to search for keywords within a specified distance of your chosen search area. Although the functionality may be replicated using Twitter's advanced search, I like this tool for the simplicity. Try it for yourself at http://nearbytweets.com and see what is happening in your area.

Twitter Polls

I believe that getting to know people using Twitter is crucial, as you will learn throughout this book. I try to know as many good people as possible, individually, but it is often useful to know the picture of these people as a group. Polls can be a great way to learn more about the people you communicate with on Twitter. Polls can help you to gain a greater understanding of mass opinion. In my experience, a quick and easy poll is a great tool, and can be extremely easy to implement. The polling tool I like to use is TwtPoll (http://twtpoll.com). There are many polling tools available, but this was the first one I used in conjunction with Twitter, and I am comfortable with it. This tool allows you to create polls by simply entering a question and a few possible answers. For example, here is a poll for you to take right now: Do you think a poll is useful for learning about friends you meet on Twitter? Answer the question and see what others think here: http://tfbtff.com/S1wIL

Data Security: Backing Up Tweets, Followers, and More

I am a big believer in data security. Two important considerations regarding data security are in keeping private things private, and avoiding data loss.

I have often expressed concerns over any Website that requires your Twitter login information, but I make an exception for backups. This is not to say they are not secure, but I am a programmer with over a decade providing wholesale Web hosting, Internet access, and programming services. I have been the geek behind the geeks for a very long time. To say the least, when you provide services to the service providers, you tend to become extremely cautious about your data. You could even call me a bit paranoid when it comes to data security, but I call it a heightened sense of awareness. The bottom line is that it would be a very simple task for somebody to create an application and store your login data. Many would never do this, but how many of them do you know and trust? With so many new Twitter applications, my errors will be on the side of caution.

I have meant to become more familiar with the Twitter API and write some handy scripts for my own use, and maybe once this book is finished I can make more time for that. In the meantime, I want to be sure I do not lose all of the data I have acquired and developed using Twitter.

A Website I have used for backing up my Twitter data, including followers, friends, favorites, tweets, and direct messages, is

Tweetake (http://tweetake.com). Using this utility, you may download a CSV (comma separated) file of your data that you may view as a spreadsheet or use in many other ways, such as importing your contacts list for use with your email application (e.g. Microsoft Outlook, Eudora, LotusNotes, Thunderbird, etcetera).

Skype

Skype did not come on the heels of Twitter, but it is a great tool that works well in conjunction with Twitter. Whether you use Twitter for business or just for fun, you can surely see the value in expanding your communication with others. I very often use Skype to enhance communication with friends I have met using Twitter. Skype provides a very simple way to create conference calls, video calls, or real-time text communications without worrying about the usage caps of Twitter.

Telephone

As above, the telephone provides a great way to expand your communications outside of Twitter. If you are using Twitter for business, you would be foolish to limit your reach to Twitter. Twitter is an amazing communication tool, but it is not the only tool in the box.

On the subject of using a telephone for communicating with people you meet using Twitter, I have a funny story for you. I had a recent instance when I tweeted a business suggestion for people to reach out each day and telephone at least five friends they met on Twitter. I got the strangest response from somebody that is presumably afraid of his own shadow. He warned his following that under no circumstance should people ever do this, and he repeated his tweets with an urgency I could not believe. He then said that nobody should ever give out their phone number, and not only implied, but directly stated that I had suggested giving out telephone numbers on Twitter. Of course, I

twitterforbusinesstwitterforfriends.com

made no such suggestion. I said that I call people, and that they call me, but I made no mention at all about telephone numbers. In fact, I would even suggest that for many people, it may be prudent to not tweet a telephone number. However, I do it without hesitation. After all, I am in business and I want all the calls I can get as long as they are productive. My toll free direct line is prominently placed on my blog, which gets a whole lot more exposure than any given message I send on Twitter.

The amazing irony of the moron ranting was that the person actually had a Website, and a business, but yet the company telephone number was not anywhere on their Website and their WHOIS data was that of a private registration. For those who do not know, the WHOIS data for a Website is the contact information for the administrator of each and every Website. It is publicly available for all to see, for every registered domain name. In fact, the rules of ICANN (Internet Corporation for Assigned Names and Numbers) require that every registry make the entire WHOIS database available for purchase. See a live WHOIS lookup for yourself at one of my favorite Websites (of course, I am the CEO of this particular company) http://yournew.com/whois.

Which one sounds safer to you? Is it the guy ranting about how unsafe using a telephone for business communications is and hides the public data about his company, or the guy who makes himself available and you can call or drop a letter in the mail and know that he will receive it?

Note that this is most important for the people doing business and using Twitter. This is not necessarily a good plan for the 13 year old tweeting to their schoolmates, or the soccer mom planning who will pick up Johnny. If you have a business and you are making it harder to grow communications with others, you are making a foolish mistake.

The most curious outcome of the paranoid guy ranting about not furthering communication beyond Twitter was that I noticed others defending what I had said. One in particular stood out, and it was a friend named Mel McKenzie (@TigerMel). When I saw her defense of reaching out beyond Twitter, I immediately went to her profile, followed the link to her blog, clicked on the "Contact Me" link, and promptly dialed her on the phone. My first word was "Boo!" and I asked if I scared her by calling. I then reassured her that I did not plan to fly to Australia to snuff out her family. We spoke at length, and we learned a lot about each other.

My conversation with Mel is not an isolated one. I have reached out by telephone to many people I met on Twitter, and I have never heard an angry word for calling. On the contrary, I often hear a very excited voice delighted to hear from me.

Twitter Implementations

In addition to many Twitter-based tools, many existing Websites have implemented automatic Twitter updates as a part of their service. They allow for you to provide your twitter login so they may update your Twitter feed when you perform actions on their service. I use these sparingly, as I addressed in the section of this book on Twitter security. Some exceptions I make are listed here, and I find them to be quite useful.

Posterous

Posterous is a very convenient and simple way to post information to the Internet by way of email or SMS text messaging. It could be said that Posterous is the simplest blogging platform ever, both in setup and updating. I use it to share personal items that do not fit in with any of my other blogs, such as how my daughter said *"Let's eat 'em"* when asked what we should do about the rabbits in our garden (http://tfbtff.com/garden).

A Posterous account may be set up to post to your Twitter account and other social networks with every update you make, while being added to your Posterous blog. When you have more than 140 characters to express, wish to share photos, videos, or other information, yet you want to keep it easy, this is a great tool. It is an easy way to begin a personal blog, and certainly the easiest to update. It will also work from nearly any phone.

There are other similar tools, and you may also implement this functionality along with the most common blogging platform, WordPress. However, if simplicity is important to you, I suggest checking out Posterous.

FriendFeed

FriendFeed is a service that allows users to consolidate multiple social networks (presently totaling 58) into a single feed. Using FriendFeed, you can be informed of updates that your friends make on other networks of their choice, providing you a bird's eye view of their activity. FriendFeed will allow you to make updates to Twitter, from the Web or may be used from most mobile phones.

If you participate in other social networks, FriendFeed should not be overlooked. I could probably make FriendFeed the topic of my next book, but for now, I will simply say that it deserves your consideration. Learn about FriendFeed at http://friendfeed.com.

Desktop Applications

There are a variety of useful Twitter applications based on the Adobe Air platform, and whether you choose Seesmic Desktop, TweetDeck, or another, you will find much greater use and enjoyment of your Twitter experience with a good desktop application. Each of these are free downloads, so you are sure to get your money's worth. I have used the tools mentioned here in my book, and more. I have drawn my opinions based on extensive use, and I want to share it with you.

Seesmic Desktop

I must give my highest regards to the team at Seesmic for their products, and their vision. This is a company that really understands social media and they develop their products by listening carefully to what others want. I have been a fan of TweetDeck in the past and I still believe that they have a fine offering, but I must credit Seesmic for how they won my loyalty.

In preparation for writing this book, I had a few questions for Seesmic and their competitor, TweetDeck. I tried many times to reach a human being at TweetDeck, but I received no communication from them. Since I had used both offerings, and I knew that Seesmic was making huge strides, I felt that I should investigate a bit further before I released my opinion to the public.

I had a very impressive experience with reaching out to Seesmic. I used Twitter to reach them, which should be no surprise. I sent out a message saying that I had tried many times to reach TweetDeck and wondered if I would have better luck with Seesmic. I was promptly greeted with a message from John Yamasaki (@jyamasaki) from Seesmic with an email address to expand on the information I would like to discuss. I was offered an interview with Seesmic founder, Loic Le Meur, which was scheduled for Monday, just two days later. At this point, Loic was a very busy man, with the launch of a major software update scheduled for the following day. He still made time for a half hour Webcast interview. Now that is dedication to social media!

During the interview, I was delighted to find that Loic is very attentive to the needs of users, and if you make a suggestion, you may just see it in their next software update. Based on our talk, I strongly suspect that although Seesmic Desktop's inclusion of additional features may differ from the release of features in other applications, Seesmic will leave no stone unturned when seeking what people want.

Aside from the fantastic service from Seesmic, I find that their product is just as good as their support. Seesmic provides the ability to navigate through information efficiently, and includes many features that I have come to appreciate. Something I really like about Seesmic Desktop is the ability to quickly and easily navigate across columns from a navigation menu on the left side. Since I have many groups and search columns, I find this to be very useful. As you read further, you will find why I believe this to be extremely important.

You may download Seesmic Desktop for yourself at http://desktop.seesmic.com and see why it is my favorite.

TweetDeck

I do not endorse many Twitter-related tools, but TweetDeck has clearly been the public favorite. TweetDeck has become the second most used Twitter tool, while the Twitter.com Website is number one.

When I started writing this book, TweetDeck was my favorite, and today it still runs a close second to Seesmic Desktop. You may download this tool at http://tweetdeck.com and judge for yourself.

Setting Up Your Desktop Application

You will have much greater enjoyment and better communications by taking these few simple steps in setting up your desktop client. People often comment on my practice of responding to every direct message and @ mention that I receive, and using a good desktop application and proper setup is the only way I can imagine doing so.

Become familiar with each button, from the settings to the retweet button, to adding friends into groups, you should try to know and understand every function. It really is not hard if you will just be patient. Once it is done, I believe you will be glad that you took this extra time.

Many users will find the default settings to be adequate, but here are some modifications I suggest that will help you to have a better experience.

API Settings

One of the first things I suggest is to set the API usage settings longer to assure that you will not reach the Twitter API limit of 150 requests per hour. If you reach the limit, you may find, as I have, that while you think you are sending updates they are not actually reaching anybody.

In Seesmic Desktop, you will click the settings button in the lower left corner, and then click on "accounts" and select the account you wish to modify. You may set the number of automated requests per hour, and also the priority of this allocation between "Tweets", "Replies", and "Direct Msgs" by moving the sliders to shorten or lengthen the refresh interval.

In TweetDeck, you will look in the upper right corner to find an icon that looks like a tiny wrench. Click on the wrench and then click on the "Twitter API" tab. You will find settings for how often TweetDeck will automatically check with Twitter for new information. The settings are categorized by "All Friends", "Replies" and "DMs".

Setting the automated refresh rates longer will reduce your API usage and also allow more time to read before the next wave of information is delivered. This can be particularly important if you follow a lot of users. If you want to refresh the information sooner, simply click on the refresh button. Seesmic Desktop allows you to refresh on a column by column basis with an icon that looks like a circulating arrow. The refresh function in TweetDeck will refresh all columns at once (a bit wasteful of API if you have many columns) and it looks like two circulating arrows and is found in the upper right corner.

Another setting that can greatly reduce API usage is a setting under the "General" tab of the TweetDeck settings. The option is listed as "Open profiles in web page". If you do not use this option, each time you click on a username it will use another API request to open the user's profile in TweetDeck. I hope to see this important option available in Seesmic Desktop soon.

Creating Groups

Even if you never make it beyond the next couple pages and you plan to throw this book on the fire, please read this next part first! If you are not using groups and saved searches to help you manage the information you gather using Twitter, you are probably missing a lot. If you are already using these tips, please be sure to share this with others so that they will have a better enjoyment of Twitter and also so they will continue to hear what you have to say.

There is a point for each of us when we simply cannot read every bit of information that comes to us. When I started using Twitter, I tried to read every single thing that everybody I followed sent. This lasted for a while, but as I found interest in so many people, it became impossible. A lot of information is great, but knowing what to do with it and making it useful is even better. Whether you follow few or many, creating groups can help you to organize your communication and be sure that you will receive communications from those you value the most.

Once you have added users into groups, as these members send messages, they will be aggregated into a separate column for that group. In my instance, I find it useful to have groups for auto racing, clients, social media, close friends, and etcetera. Consider what groups make sense for you, and try this tip. I think you will like what you find.

In order to create a group with Seesmic Desktop, look to the navigation column on the left side of the page. Click on the "+" icon beside "Userlists", and you will be asked to enter a name for the list. Enter a name for your new list. Once the list is created, you can add users to the list easily. Each message you receive will have an icon that looks like a gear when you place your cursor over the user's icon. Click on the gear icon and select "Add to userlist" and then select the list you wish to add the user into.

Using TweetDeck, click the "Group" icon at the top of TweetDeck, choose a name for the group, and then select the members from a list. You may also add users to your groups from the "Other actions" icon when you point to their icon.

Saved Searches

Both Seesmic Desktop and TweetDeck make it simple to create search functions for topics that interest you. This is particularly useful if you are protecting or cultivating a brand. For example, I will monitor a search for "tfbtff" because this is the hashtag I will use for the marketing and support of this book. Seesmic includes a search form in the upper right corner and TweetDeck has a small magnifying glass shaped icon at the top of the window. When you enter a search term, a new column is created that will be updated with near real-time results for your query. This is a very powerful means to stay abreast of information of your choice. It can also be a great way to find other Twitter users with similar interests or within a geographic area. I also use searches for usernames of people whom I wish to follow very closely. This allows me to keep close contact with them and also follow their conversations with others. A good use for this is when a friend is sick, getting married, having a baby, or just downright interesting. Please remember that when you enter a search, you can use any of the advanced search operators that I listed in the section about "Twitter Tools". This means you can be very specific about the information you receive in a saved search.

Multiple Twitter Accounts

If you have a need to manage multiple Twitter accounts, a good desktop client will save you a lot of time. Seesmic Desktop and TweetDeck both provide for managing multiple accounts. The most common cause for this is having separate work and personal accounts. With these tools, you can manage and monitor each of them from a single place. An extreme instance of this was given by Loic Le Meur of Seesmic during our interview. He told of a person who uses Seesmic Desktop to manage 25 separate Twitter accounts for his clients. This is a good example of the efficiency afforded by a good desktop application. This is another instance where I give much credit to Seesmic for making it easy to manage and navigate many columns.

Mobile Tweeting

The ability to update Twitter from a mobile device has arguably changed the way news is reported. There have been many instances when the first reports of major news have come from a Twitter user with a cellular phone.

An early misconception I had about Twitter was that my primary use of the service would be from a phone. I believe that something in the earlier presentation of Twitter led me to think of it as a mobile oriented service. This is a feature of Twitter, but I do not see it as the main feature. It is certainly a great facet of the service. Many people update regularly from a phone, and I believe you should become familiar with it. Even if you do not intend to use Twitter from your phone, I would suggest registering your phone with Twitter so you can keep the option available.

There have been many applications developed for mobile devices, and they change all the time. Since there are many different types of phones and new applications developed regularly, I first want to discuss features that will work from any phone with SMS text messaging capabilities.

Register Your Device

In order to use SMS text messaging to post to your Twitter account and receive messages from other users, you must first register your phone with Twitter. This is a straightforward process, and it is a free service from Twitter. Of course, there is the obligatory disclaimer that standard text messaging rates apply, contact your cellular carrier for details.

Registering your device starts by clicking on "Settings" while logged into your Twitter account. From there, click on devices, and follow the simple instructions for verifying your phone. Once this is set up, you can select what messages you wish to receive, and also a schedule.

SMS Commands

When you use Twitter with SMS text messages, the first thing to note is that your message will always go to the same number. I know, this may seem elementary to many people, but for those unfamiliar with SMS, you should note the number you will use for the commands I will discuss here. The number to use based on your location is as follows:

USA - send commands to 40404
Canada - send commands to 21212
UK - send commands to +44 7624 801423
UK (Vodafone) - send commands to 86444
Sweden - send commands to +46 737 494222
Germany - send commands to +49 17 6888 50505
Anywhere else - send commands to +44 7624 801423

Now you are ready to send a message directly to your Twitter feed, to individual friends, receive updates, and more. Here are the commands you will find the most useful.

If you want to send a basic update, simply create a new text message to the number above and say whatever you like. Since your phone is registered, Twitter will recognize you as the sender and it will be added to your timeline. Do not forget that updates are capped at 140 characters. It is as simple as that, but there is a lot more you can do, too.

If you want to send a direct message privately to somebody who follows your Twitter feed, you must prefix the message with the letter "d", a space, and then their username. (Please note that there is no "@" symbol here.) This is the same process as sending an update, except that it is prefixed with a recipient. For example, if you begin a message with "d murnahan" the message will be sent directly to me.

If you want to send a message to somebody publicly, simply be sure that the message contains their username with the "@" symbol such as "@murnahan". Note: There is not a space between the "@" and the username, as there is a space with a direct message.

You may turn mobile updates on and off from the Twitter Website, but what if you are not at your computer? You can also do this using SMS text messages from your phone. These commands are sent to the same number, just like sending updates, but will not appear in your Twitter timeline. Here are some commands to help you turn updates on and off, and more.

OFF – If you send just the word "OFF", it will turn off all notifications. If you are walking into a meeting and cannot be bothered, you may also send the word "STOP" and it will stop messages immediately.

ON username – You may turn notifications on for specific users by text messaging the word "ON" followed by a space and then a username. In reverse, sending "OFF" followed by a space and then the username will turn notifications from that user off. Examples of this are "ON murnahan" or "OFF murnahan".

GET username – This will retrieve the most recent post from a user. If you want to know what I have posted most recently, just message GET murnahan and it will be automatically messaged back to you.

FOLLOW username - You may follow a user on Twitter by messaging "FOLLOW username". I think you will find this command particularly valuable if you are not already following my Twitter feed. Simply send "FOLLOW murnahan" and find out what I am up to. Be sure to also send me a message "@murnahan" and make an introduction.

Mobile Twitter Applications

In order to get the very best use from Twitter on your mobile device, you will need a good Twitter application. Of course, you can access Twitter's mobile-optimized Website on any phone with a Web browser at http://m.twitter.com, but similar to using Twitter from your desktop, there are many other options. The best applications are developed by third parties outside of Twitter, and it seems that many developers want a piece of this pie.

Meet Mashable

I could write a whole book about the many Twitter applications available for mobile devices, but they change often, and there are many different mobile platforms. I believe that you will be better served by an introduction to Mashable.com than a lengthy explanation of these many options. Mashable.com is a highly respected social media blog, and the Mashable team has published many excellent articles on this topic. This is a site worth your attention, and not just for Twitter. Here is the URL for information on Twitter applications: http://tfbtff.com/531Zl

Tweeting With Strangers?

Are we tweeting with strangers? This question surely makes sense to a lot of people. You may wonder how to meet people with Twitter or how a person can possibly ever feel they have developed friendships and alliances using the Internet. Can you imagine developing meaningful personal and business relationships using what has sometimes even been so misunderstood to be called an "Update Service"? It is often the case that a proper answer is far lengthier than the question. Fortunately, that is why you have this book. Just to ease the burning question of whether we are just tweeting with strangers, I will give you the simple answer, which is "no".

A more in-depth look at the question at hand will reveal that many great relationships begin with a greeting far shorter than the 140 character limitation of Twitter. Think about how many times you met somebody and the first thing you said to them was simply "Hi". Consider how you met your dearest friends, or even your spouse. If you can recall the very first communication, it likely all began with one single sentence. Sure, you may have had a great conversation, but when it started, you said something, and then you listened to what they had to say. I trust that you can see where I am going with this.

Is there a right or wrong way to use Twitter?

There are some things that you would surely not wish to reveal in public, and there are some things that you will share with some friends but not with others. There are clearly things that are against the rules of Twitter, and of society for that matter. However, when it comes to Twitter, what I want to make abundantly clear is that the people using Twitter truly are as unique as those you would meet in a shopping mall, restaurant, your children's school, or anywhere else. You can meet many types, and each has a unique style of using Twitter. Being yourself is far better than trying to conform to how anybody may tell you to use the service. Be real, be yourself, and rest assured that there will be many others who will like you just the way you are.

Overcoming Shyness, Hesitation, and Laziness

If I could help eradicate shyness, hesitation, and laziness just by writing a book, we would really have something special there, right? If this book even makes a significant impact in those areas, I will clearly have a better 2010 racing season than I ever expected. You are surely wondering what the heck this has to do with racing, but pay close attention! You see, you just learned more about me. Yes, I am a race car driver. I instruct high performance driving and racing skills for Porsche Club of America, Audi Club North America, BMW Club of America, race tracks, and several other driving courses including Street Survival (http://streetsurvival.org) for young drivers. It is my burning desire to race cars full time, while considering the stuff that pays me to be my hobby. I also love to help others race better, faster, and safer. I am very accomplished in this area. If you were receiving my tweets, you may have even seen me send a tweet of one of my on-track videos or show my Cop Magnet race Webcast at http://copmagnet.com. So what does this have to do with social media and Twitter, you must ask? It has everything to do with listening to people and realizing that you can meet and get to know people much more easily when you know something about them.

Enter a Warm Conversation

When entering a warm conversation, you are not making cold connections with people you know nothing about. Not unlike any in-person social gathering, if you observe what is around you, there are many ways to tactfully join into a conversation. If I tweet about it, then it is clearly not meant to be private. If somebody tweets about auto racing and you like the sport, send them a message. It could be as simple as saying "I see that you teach teen drivers. Do you know a good school in Atlanta?" It is very non-threatening, and how do you really expect somebody to answer that? I can tell you that it will normally be met with a very friendly response. Twitter users are there to meet other people, so don't be shy.

Hesitation: The Real Enemy is You!

When I think of hesitation, I picture the person who simply waits for good things to fall into their lap. I receive many telephone calls from people who say they have a great product or service they want to bring to market, but then never follow through with a plan. When it begins to look like an effort is required, many people will hesitate. I used to take a lot more time following up with these people, but I found that they will generally hesitate for at least six to twelve months before they even take the slightest action. Sure, you could assume that a simple call from me could prompt action on their part, but at what cost? If I spend my time chasing the people who talk more than act, I am reducing myself little by little into their mediocrity. So many of them will tell me about how their plans changed, their committee needs to meet again, or that they are still excited to work with me. The bottom line is that in the time they hesitate, the highest cost is accruing right under their nose. This cost is that of lost opportunity. When you hesitate and talk about thinking about considering possibly one day doing something to improve your network of friends and business relations, you are only hurting yourself.

Laziness: 3 Simple Steps

On the topic of laziness, it is easy to spot. Maybe I could come up with a gentler word for it, but so many of the things you wish you could achieve with social media, yet have not, come down to laziness. You can say that you just don't have the time for it, but that is quite often simply an excuse. You have the same amount of time the rest of us have, but an important question is in whether you are using that time doing the right things. Did you make an effort to reach out and meet somebody new today? Have you taken steps today to create a stronger and wiser approach to achieving your social media goals tomorrow? If not, how could you possibly expect positive results?

I am going to give you three daily tasks to perform that, when done consistently over time, will curb your laziness and give you more of what you want.

Step 1. Meet somebody new today on Twitter! If you cannot think of anything to say to them, keep looking until you find somebody who interests you more. You will see funny stuff on Twitter all the time. Maybe you can just comment on their wit. Whatever it is that sparked your interest, let them know. Something, anything, just go meet somebody! Start a dialog, even if it as simple as "Hi", but even better, find out to whom you are saying "Hi".

Step 2. Communicate with the new person in Step 1 every day for at least a week. If you have chosen somebody you found interesting, you will want to do this anyway. Wish them a good day, ask if they are still _____, find out how their kid's school program went ... just say something. I am not saying that you should force out something that is not genuine, I am saying that if you want to get to know somebody, you must make an effort.

Step 3. Repeat Step 1 and Step 2 every day for 30 days! At the end of 30 days, you will have made some new friends and you will be well on your way with a new and positive habit.

If you are disappointed with your social networking and social media marketing efforts, laziness is likely the first place to look. This is true with many things missing in your life. Are you making every effort to reach out and use the tools at your disposal? Do you ever find yourself thinking that you just don't have the time to keep trying something that does not show you an immediate benefit? Follow these three simple steps. If it does not benefit you the way you expected, the worst scenario is that it still helped.

If you will spend the time to know and care about others, you will build relationships based on far more than the few words that you see in a tweet. The basic principle here is to open your eyes and really see what people are telling you. The question of "Who are these Twitter people?" will soon fade away.

The Following Frenzy

Upon first using Twitter, it can be a bit confusing by all of its simplicity. An early step to finding usefulness in Twitter is to meet others and follow their Twitter feed. Once a conversation begins, and you find a few people to follow, it is easier to understand the value of Twitter. With a little experience, it becomes obvious why Twitter is talked about on every television and radio station, and why people claim it to be the future of media.

Something you will find quite abundant among Twitter users is the frenzy to reach more followers. The allure of the initial value of finding people to communicate with seems to act like heroin for some people. It becomes a quest that can overshadow the best use of Twitter, and can make an otherwise productive user into a monster. I implore you to be reminded that too much of a good thing can destroy the goodness.

A question you must ask yourself before falling into this trap is "what is the value of a Twitter follower?" Is it going to make you the star of your high school football team? Will it make you prettier so you have a better shot in the cheerleader tryouts? Will you use that incredible popularity to turn it all into millions of dollars?

It seems that having enough followers to fill a stadium provides some great sense of value, but unless you know what it is, or why you try so hard at increasing your fan base, perhaps it is best to slow the pace a bit and determine what you are after. This answer will be different for different people, but it is something I believe you should put your finger on before you engage in the practice of seeking the largest follower base.

We all know that giving even the slightest glimpse of what you do to earn a living is totally taboo in social media. After all, this is social media folks ... nobody here actually has a job or works for a living. If you have not already learned this lesson, you must not be following enough self-proclaimed social media experts or jealous antagonists who will try to sneak in their ads under the radar in hopes that people will swarm their Website and place orders without knowing that they are actually buying from somebody they met through social media. O.K., you've got a whiff of my sarcasm, right?

The long and the short of it is that without either a desperate need for personal validation or a legitimate and useful business purpose, there should be no need to have a *squillion* followers.

The Nerdy Kid From Math Class is Your Friend

It is clear enough that many people using Twitter are doing so with the simple intent of having fun, meeting people, and enjoying learning the many wonders this smaller world offers. These are the ones so many of us like, because they pose little threat, and they are never trying to sell us something or teaching us how to get thousands of Twitter followers. They are also often the people without a massive base of followers, and not because they are not just as valued, but because they do not choose to participate in the popularity contest. Perhaps, in following my earlier comments regarding the star of the football team or cheerleader, this is the dork with the thick glasses and high-water pants. In this case, I feel right at home, because those are the kids I defended in school. Those are the kids I still talk to today. Those are the kids who somehow seemed real to me. Later, I found that they were also often the ones who avoided drug overdoses, divorce, jail, and many of the other things I watched the popular kids enjoy as their circle of friends crumbled and their popularity dwindled.

This nerdy kid from math class with only an itty-bitty follower count can be your best friend. I certainly find myself having meaningful chats with this nerd. The nerd knows me, and they know that I know them. They have earned my attention as I have theirs. They care what I have to say, and I care what they have to say. We may not always agree, but then, I disagreed with my wife once (but only once)!

This nerdy kid I write about may have only heard me because there was nobody else talking to them, or because they were only listening to people who cared enough to reach out and be their friend. What may shock you is that the same nerdy kid may also decide that once they know you and your intentions, they want to hire you, buy from you, tell others about you, or leave their estate in Hawaii to you.

This nerdy kid is the majority. This nerdy kid has my respect and my loyalty.

How "Cool People" Have Many Followers

We all see the "cool" people with their amazing number of Twitter followers, but don't you sometimes wonder why so many people are following them? Their value in the Twitter community is not always in proportion with their number of followers. I can tell you the top four ways it happens:

1. They repeatedly follow as many people as Twitter will allow, because the vast majority of people will return the follow.

2. They follow everybody who follows them, but then promptly unfollow anybody not following them. This way, Twitter will allow them to follow more (see number 1).

3. They represent a value to other Twitter users, and people want to hear what they have to say.

4. They have a television, radio, print, Internet, or other history that draws people to follow them on Twitter.

With regard to number one, this is something I personally do not find to be of great value. However, some people that I actually kind of like to follow are engaging in this on a regular basis. This is their way, and I do not hold a grudge, but I also know that there are a very miniscule number of people doing this who will ever deliver the value which their popularity would seemingly represent. This is a tactic used by many Twitter users ... perhaps even you. It also yields the lowest percentage of users reading what you tweet, responding to it, or having any meaningful interaction with you.

Number two is often used in combination with number one. I have wanted to be the guy who follows all of the people who were so "interested" that they would follow my tweets. I still test this on occasion, but it has become just about as useless as number one. I recently tested this by blindly re-following roughly 1,000 people who were following my tweets. To put it mildly, I will say that I found many opportunities to get thousands of Twitter followers, buy more real estate, read more SEO blogs, and join in more games. Yet, I still cannot determine a single user from this test group with which I have held a useful conversation, read an interesting message from, or grown my business. The amazing and wonderful result was that it did grow my follower count! So what gives? Well, number one and number two are closely joined. It seems that when I return-follow each user following my Twitter feed, my following grows massively. This is perhaps largely why there are many users following my Twitter feed. If I did not follow them, many would not stick around long enough to find whether I hold any value or not. They simply use a combination of steps one and two. I have commonly re-followed most people who follow my Twitter feed, which I believe is useful so that they may direct message me if they should choose. However, even when I follow them, they should know that I do not see everything they send publicly ... nobody following a massive number of people will. If they want me to see everything they say, they will engage in conversation, become my friend, be added to one of my Seesmic Desktop groups, earn their way to my "Follow Close" group, and then I will read every last word they ever tweet. Short of that, just having me follow them is less useful than putting "@murnahan" in the tweet, because I read 100% of those!

On to number three from the list, I really hope and believe that I am one of these people. I share what I think, I tweet some funny, entertaining, and useful stuff. I engage with anybody who will talk to me, and I am available outside Twitter at any time of night or day. I am the guy who really does care to hear from you ... try me!

Number four ... well; I know some of these, too. One that comes to mind is Ashton Kutcher (@aplusk). People have called him "The King of Twitter", and while nearing three million followers, he boasts a higher follower count than any other user. I criticized his campaign for followers, and said I was not going to join the crowd. I do follow Ashton now, but that all started when Ashton followed my Twitter feed, and I wanted to be sure he could return message me directly after I sent him a direct message. Since that time, he has been pretty respectful toward me. We are not close buddies, but we do direct message back and forth once in a while. I sometimes suspect that he is actually seeing a lot more than people think he is. I have also heard through the grapevine that M. C. Hammer (@mchammer) is also a bit more engaged than one may anticipate. However, we can generally expect that this group will not show up for your next family reunion, holiday party, or tweetup.

The Truth About Follower Numbers

After responding to a blog post that criticized the hunt for huge numbers of followers, I read a response to my reply on that blog that said "If you need to read a book about Twitter, you're doing it wrong". Of course, to this, I found a great deal of inaccuracy.

Many people will unsuspectingly fall into a pack mentality and do what others do. It is very normal to assume that the masses are right, and that joining in their bad habits is just fine. When this happens with a frenzy to have more followers, the whole pack suffers with a sort of deafness. It becomes increasingly harder to hear over the static. If I can help to break you away from this pack, and to find a more useful approach, I think reading a book about Twitter is perhaps not so bad after all.

Perhaps the greatest value of social media is in gaining a greater understanding of those around us. Whether we are making friends or selling weight loss products and Viagra, we can all share knowledge. I realize that you will find value as you see fit. I can only hope that you will see the truth about this follower frenzy as I have described it. This is the truth according to Mark: It is better to have 100 friends who will say something nice about you than any number of followers. When you hold a value that others can see and appreciate, the rest just comes naturally.

Three Great Ways to Meet Twitter Users

I will show you some ways that I meet other Twitter users, or as many call them, "tweeps". I believe that these ways of finding others are far more beneficial than randomly following everybody you can follow.

Meet by Introduction

Have you ever noticed that in your circle of friends, you have met some great people by introduction from another friend? It works this way with Twitter, too. I meet many people because somebody thought I should meet their friend. This is a basic principle of networking. I am introduced many times each day, and I always try to introduce people that I believe may have something in common or who may benefit from knowing each other. It is also a fine idea to ask friends to introduce you to others who share an interest. I always try to keep an ear out for somebody hoping to meet people in a given category. I remember not so long ago being asked if I knew any people using Twitter in Australia. I immediately made efforts to connect them with many friends I have made in Australia.

Mutual Friends

If you observe a friend carrying on a conversation with somebody, take a look at who they find interesting. You may not have a lot in common, but if your friend thinks highly of them or has good conversations with them, it may be a good indication that you will as well. I like to find people who have something to say. I follow a lot of Twitter users who do not have a lot to say, but the active ones are generally much more fun. I told you about Twitter Analyzer (http://twitteranalyzer.com), so be sure that you don't ignore my tip. You can use this tool to see the people your friends communicate with the most, which can be a great way to meet actively engaged friends. As you do this, you will soon find that you have developed a social circle with many mutual friends.

@ Mentions

If somebody tweets something with your username, it will come to you as a mention. This does not mean that it was directly to you, but if somebody is saying something about you, don't you want to know why? It happens that this was how Ashton Kutcher learned of my criticism of his race against CNN for one million Twitter followers. I will discuss that in another chapter.

Whether somebody is retweeting something by you or just mentioning you to a friend, you will do well to see who it is and find whether they are somebody you will want to know. If they are retweeting you, there are two benefits to following them. First, you know that they understand how to retweet content, which is a huge asset. Secondly, you know they have already heard your name and appreciate your content. In many instances, I will find that somebody who retweets my content is already following my tweets, so I certainly want to hear what they have to say.

Targeting Using Twitter

Reaching a specific audience is extremely important for marketing, and also for friendship. Finding others with specific interests and in specific locations has never been easier. Twitter provides great benefit to those seeking a targeted group, if you know how to use it. Here are some tips for reaching the people you seek.

Geotargeting Using Twitter

Using Twitter advanced search features, you can find what people are tweeting about by location. If you took my tip on using a desktop client such as Seesmic Desktop or TweetDeck, try entering a search like the following:

near:KansasCity within:10mi grilling OR cooking ?

This search will return results for tweets within 10 miles of Kansas City with the words grilling or cooking and asking a question. If you were operating a cooking school, cookware store, or grocery, this may give you a great opportunity to help somebody with their cooking questions. If you run a tire shop, you could search for flat tires, or a restaurant may search for hungry OR dinner. You get the idea, right?

People like helpful people, and they may just remember it the next time they need your help. When you enter this in your desktop client, you will be constantly updated with new results.

Another handy tool is Nearby Tweets (http://nearbytweets.com), which provides a similar function from a Web-based tool. With all of the many available tools, and let's not forget the advanced search functions at Twitter.com (http://search.twitter.com/advanced), any concerns of reaching people based on geography should be overcome right now.

Demographic Targeting Using Twitter

Similar to geotargeting, it should be simple for you to reach any target demographic using Twitter. In my experience, searching specific terms to find people tweeting about a topic will lead you to an even more focused audience than expected. Using multiple search terms and being creative with your searches will take some practice, but you can quickly find just what you are looking for. It will also allow you to find people who are talking about a topic right now, meaning that you are not reaching a cold audience.

Although much is still speculative as to the overall demographics of Twitter users, largely due to the extreme growth figures, you may find very interesting estimates of Twitter user demographics by visiting Quantcast.com (http://quantcast.com/twitter.com).

Tweetup: A Twitter Meet-Up

Tweetup is a term for a meet-up of Twitter users. It is a way to bring people together and provide proof that these are actually real people who have real lives, and really are your friends or colleagues. You can have a tweetup for business, schoolwork, sports, or anything else you may choose. The topic you select is up to you.

When asking my wife, Peggy (@pegmu) about this subject, I mentioned that I had been invited to a tweetup just today. As funny as it may sound, I actually heard about it for the first time because a friend sent me a Facebook message about it. This further proves the importance of diversity in your communications tools.

I told Peggy the tweetup I learned about today was at a restaurant in my town that we have not visited. I asked her if it was possibly the place where we once attended a Chamber of Commerce meeting. This question started a great expansion of the topic, and we started reminiscing about once upon a time when we thought the Chamber of Commerce may be a good idea. In order to show a contrast in the old school and new school methods, I will tell you a story about my Chamber of Commerce experience. Then I will let you draw your opinion on the usefulness of a tweetup.

Some things we recalled about the Chamber meetings we attended were that it seemed impersonal, and full of a bunch of braggarts wielding business cards and brochures. I am sure that there are some Chamber of Commerce meetings where there is meaningful work being done, and probably a lot of friends being made. I am just sharing my experience, which introduced me to a lot of arrogance, ignorance, and elbow-rubbing with the good old boys club. It felt very uninviting, cold, and it lacked a lot for me. It seemed like a group of "if you buy some of my junk, I will buy some of your junk, and together we will all get rich."

A sign of just how impersonal this group was to me came after I decided that I would not renew my membership. Apparently there was a remnant of my past membership left on one of my Websites, in the form of a logo and link to their Website. I received an email message one day that quite sternly reminded me that I am not a member, and that I must immediately remove the logo. This person not only pointed out I had some old content needing updated, but she also showed me just how ignorant the Chamber approach was in dealing with people. If it was my Chamber to influence, the first reaction would have been to ask if there was a reason for not renewing, and if there was anything the Chamber could do to make the organization more useful. Now, mind you, I did not leave on bad terms or anything. I just faded away and decided it was not for me.

Surely I sound like I am not a big fan of this group, but I have also experienced much of the same with other Chamber of Commerce groups in other areas. I hope that your experiences are much different, but this represents my impression of the old school. Now allow me to give you a simple view of how a tweetup is much different.

As I mentioned, a tweetup can have a topic, even if the topic is just to have somebody familiar to meet for lunch. Since I am comparing it to a Chamber of Commerce meet up, we will use a business scenario. If you are meeting with other Twitter users, it is likely that you have relatively frequent communications with them. For most Twitter users, I believe they will use Twitter more often than they will pick up a Chamber of Commerce directory and start dialing somebody on the phone. Certainly, this is not always true, but as your use of Twitter grows, I believe it is likely that you will find this to be the case.

Chamber of Commerce holds somewhat frequent meetings and encourages communication. However, the methods are commonly not at all as engaging as that which is found with social media. Using social media tools such as Twitter, it is much easier to stay in touch on a daily basis, and not just when you have the time to make a call or write an email message. This is largely why social media has been said to have eclipsed email in usage.

With Twitter, as with other social media tools, you can create groups. One group may be your closest friends, another may be your knitting group, and one may be business connections. If you use Seesmic Desktop or TweetDeck, you can easily create groups that can allow you to monitor when anybody in your group has something to say. You may also choose to use tools like hashtags or Nearby Tweets (http://nearbytweets.com) to follow topics.

Considering the capabilities of focusing or broadening the groups you choose to have a tweetup with, surely you can see how this can be a benefit. Once you add in the inherent regularity of communication, it should be clear just how much the new school has taken over.

Types of Twitter Users

Social media is often a very personal matter to those who embrace it. Because social media is very personal, there is a lot of room for growing or destroying relationships. Some will say that because of the inherent lack of vocal inflection, facial expression, body language, etcetera, that there is much missed in the communication. This is true, but it also emphasizes the communication that is used, which is the use of words.

I will outline some of the things you will frequently find in others when exposed to Twitter in-depth. Many users will have characteristics of more than one of these types. Some of these are great, and some are not so great, but you will do well to be prepared for each. This is not a rant. I simply wish to share my opinion with you in hopes that you will recognize the value of each of these traits I often find using Twitter.

The Expert

On the heels of the Twitter explosion comes The Expert Twitter user. It seems that you will hear more complaining about the Twitter Experts than you will actually hear from them. The Expert is the one selling a book, service, or consulting about Twitter or other social media.

Many of the so-called Twitter experts often preach growing your following so you have more people to reach with your message.

There has been a huge surge in the promotion of programs to gain more Twitter followers, and it seems that the majority of people, including me, are unhappy about this.

Note that, although on the surface, it may seem that I fall into the category of The Expert; you will find that I am quite unlike these folks. Sure, I am selling a book and I am a social media consultant, but my method centers around building relationships rather than selling or advertising to your following. It is far more productive to have many friends who will do your advertising for you by mentioning you when somebody they know may need what you offer. This is relationship marketing and relevance which I encourage.

The Experts is frequently criticized for many reasons. The most common complaint I hear is based on the presumption that if you are an actual expert, it will show without having to tell people. Second to this primary complaint is that people are really tired of those who value them as a number rather than as a person.

The People Pleaser

This is the type who will say anything you want to hear. They often do this appended with words like "hugs" or "kiss". They are so sweet that they may give you a toothache. The People Pleaser focuses on trying to make everybody feel warm and fuzzy. If you want friends who are too afraid to tell you the truth, you may enjoy The People Pleaser. I have often found them to be the same person who will nastily try to rip your head off if you ever call them to the mat for the fake they really are. They are the defenders of all evil, and superheroes for those with a suffering self-esteem. If you want to get a real lesson in making friends, see The People Pleaser for what they are and don't take their words too deeply to heart. A friend will give you the truth when you ask if your breath is bad or if those pants make your butt look big. Perhaps you can tell that I do not categorize myself here. If your butt looks big, I will be the first to give it to you straight.

The Complainer

The Complainer can be categorized in many ways, but they are always finding something to complain about. Whether it is the Twitter service, the advertising some people send out begging for attention, or The Expert, they will complain about it.

I think perhaps the most The Complainer has to live for is to call somebody out for self-promoting. What is funny about this is that we are all promoting something, whether it is an interesting article, a funny joke, or a business. It could be said that simply using Twitter it is a form of self-promotion. After all, we all want to be seen as useful, funny, or friendly. If you promote your friendliness, it is still a promotion of yourself. An important point to remember is that if a Complainer hassles you, they have chosen to hear you, so it really is silly of them to complain.

An example of The Complainer is the joker who impersonates my name and tries to warn others that I am a spammer. Ironically, I have been very outspoken about sending useless advertisements using Twitter. This is not so much because it annoys me, because in fact, unless it is sent in the form of @murnahan from somebody that I am not following, it was solicited when I chose to follow their Twitter feed.

The primary reason I would consider repetitive advertising on Twitter to be an annoyance is because it often shows that the person sending the message clearly does not understand a more productive way to reach people. They need my book!

Now this is not to say you should never advertise. It is your Twitter account to do with as you like. Just try to remember that similar to when you sit down at your television to watch a program, the commercials are small clips between the programs you tuned in to watch. You may also wish to consider that people pay extra to watch television with no commercials.

It is easy for many people to fall into the role of The Complainer. If you hear complaints enough, you may find yourself jumping on the bandwagon. I would suggest keeping your chin up, keeping things positive, and avoid being tuned out for being The Complainer.

The Antagonist

Like The Complainer, The Antagonist often seeks out a reason to complain. The reason this one gets a separate category is the intensity and focused nature of their attacks. I think of this person much like the one who will lurk in a Web forum or IRC chat, waiting for somebody to use the wrong grammar or make a typographical error, and then pounce. It seems that Twitter brings a new sophistication to antagonizing people. I hoped I was alone, but during a Webcast on the topic of antagonists, I asked others if they ever felt antagonized and what they thought of it. Apparently I was in good company and I found that others were very tired of The Antagonist. During this Webcast, I was not surprised to find that one or more of my antagonists were present. One of them sent me multiple direct messages on Twitter. One of the messages read, "You are the one being hateful, how dare you talk about me like that. You are being ugly by bad mouthing me." The interesting part was that I was using an example set by this person, but I kept their identity confidential. I suppose what this individual may have meant was that I was badmouthing them as a group, and they had a need to defend their position of strength and value.

Beware, The Antagonist runs in packs folks! In your life, I am certain that you have seen the pack mentality of bullies. You can find it on any playground, in many workplaces, and certainly on

Twitter. You will often find the antagonist messaging their friends seeking support for their case.

One Antagonist that favors me verges on what I would call a stalker. This guy repeatedly reveals that he keeps a close watch on me. In an extreme instance of this, he sent a Twitter update to announce what a failure I am.

In the interest of teaching a lesson and also extinguishing any flame this antagonist had to burn, I will offer you a story. This is intended to illustrate that an antagonist sometimes has little back story, and shows their ignorance when they criticize what they do not know. This is my story, and it is an example of how to take away the negativity that bullies thrive on.

I do not suppose anybody using Twitter has been completely immune to economic recession by this point. You may be doing just fine and having the best economic year of your life, but the fact remains that it has touched you in some way. It touched me when a large corporation began laying off employees and ceasing services which provided for a good portion of my corporation's earnings. I began to see it in 2007, but I vowed to my clients to do my best to be the last man standing. By mid-2008 the layoffs and closures started with one, and then another, and all of the sudden, it was clear that the landscape was changing dramatically. Since early 2001 I have been the CEO of a company that provides wholesale Internet services to Internet service providers. We built a very successful company providing wholesale Web hosting, domain registrations, Internet access, Website development, Internet marketing, and more. Since we have enormous purchasing power, we were able to sell the services of many regional and nationwide dial-up Internet networks on a wholesale basis at rates that a smaller company could not receive directly from these networks. We provided the means for dial-up service providers to sell most of their equipment and eliminate up to 70 percent of their technology expense by using our managed modem services. I was providing a valuable service, I helped companies, and I helped many people keep their jobs.

The downside of keeping services running and assisting others to find replacement services was that in the end, I had to cut my own salary. While trying to mitigate others' losses, my income was decreased on a monthly basis by approximately two to three times the average American's annual salary. Oh sure, it is easy to ask why was I getting paid that much in the first place. Go ahead, antagonize, but I sacrificed my own security and very nearly lost much of what I worked very hard to achieve in attempts to minimize the losses of others. I had clients to reimburse, and extended contracts to buy out. I could have walked away, but I did what my integrity dictated. Some will call that bad business sense, and that is fine. I was criticized for having two new Corvettes, a custom chopper worth more than my first two homes combined, Picasso hanging on my walls, and etcetera. The bottom line was that all in all, I was doing a good service for a great many people, and although it may sound decadent, my lifestyle was based on only a very small portion of my earnings, which I gladly gave back when it made sense to me. Can the antagonist make that statement? I doubt that. So when The Antagonist wants to call me out for the failure I am, bring it on.

My response to this particular knuckle-dragger was to publicly respond by saying "It has been blogged. Some called it dumb, but most called it integrity." Now if you wish to criticize a man who dropped out of school at 15 years of age, has operated successful companies since 16, and done so, with integrity before gain, let's dance!

One more note on this: Although I have done much to repair my business, I should mention that this book was written *with* money and not *for* money. Sure, I would love if this book sells millions of copies, but that is a huge long shot. I wrote this because I believe in Twitter, and I know it can be monumental in your life. If you think I wrote this to spread ignorance or to deceive, take a look at the royalties of an author and then explain how I did this for the money.

I hope that what you gain from this is that if you let this person run with their torch, it may eventually burn you. Take the fire away from them. Sometimes it is as simple as saying "I regret that you took offense to this. That was not my intent." Even more often, it is best to simply ignore them and understand that the power to hurt others is often their greatest reward. When you do not give them your oxygen, their fire goes out.

This all begs the question of how many people may not achieve their goals because they took something like this to heart. Please do not let this happen to you.

The Unengaged

This is the most common person you will find on Twitter. I do not mean that they are not engaged with others, but simply that they are not engaged with you. This is somebody you want to know. Perhaps they followed you for a reason, or you followed them for a reason. Rather than simply let it be a stagnant relationship of followers, send them a message. A good practice is to go through your list each day and choose a handful of these and find out something about them. Read their tweets, read their biography, check out their Website. Once you know something about them, say hello and wish them a nice day. Most people use Twitter to meet others, so this should not be so hard to do. Try this each day until it becomes a habit, and you may be surprised who you meet.

The Tweetalots

The Tweetalot is the user that frustrates you because they tweet a lot. This one frustrated me in the beginning. I hear people talk about them, tweet about them, and blog about them. This is the power-user who has a lot to say, and it may even be the case that they have a lot of interesting things to share. You may say that they clutter your Twitter stream. You may even unfollow some of them, because you feel you simply cannot keep up.

This is largely only frustrating to you because you are new at this and have not yet realized how to handle it. Maybe you even chose to receive updates to a mobile device by text messages each time they update. This was the case with me once. I subscribed by mobile device to the Twitter feed of a local news affiliate. I quickly became frustrated because they sent too many updates. That was when I thought Twitter's best use was to be instantly updated every time each person I followed had something to say. This was not a good approach. Today I only follow one user from which I receive mobile updates, and that is @pegmu, my wife, Peggy. I would not even allow that if she was a Tweetalot.

Once you become more familiar with Twitter (and if you take my tip on the use of a good desktop application), you may come to enjoy the Tweetalot. If you are using Twitter efficiently, you can overcome all of the frustrations of these users, and they will blend in just fine. If you think somebody is sending too many tweets, just imagine what happens if you follow over 15,000 Twitter users like I do.

I am a Tweetalot, and I have many friends who are as well. They are very active users, and you will find them online a lot. They can be a great asset to your knowledge and love of Twitter.

The Friend and Supporter

I thought this would be an easy topic, but it is not as simple as I once thought. Friendships take work. If it is a true friend, the relationship will not be based on fragility, but rather flexibility and strength. Twitter offers much opportunity to meet friends. When you find one of these people, take good care of the friendship and do not take it for granted.

Be Yourself: Create a Personal Brand

Like moving back in time, the importance of bringing personality and personal responsibility into the marketplace has grown with social media. People want to do business with people. They want people they can know, trust, and hold accountable to respond to their needs. Many people can relate to this with their insurance agent or their doctor, but it reaches into every industry.

After sending out a message including a link to an article on the telephone provider Sprint, it was not very long before a representative reached out to me with a message that indicated Sprint cares about its customers. That is not exactly what the message said, but it showed that a person was in there somewhere, and that he took personal responsibility in helping me if I should ever need assistance with my Sprint account. The user who reached out to me was @JGoldsborough, and not only did he make an impression on behalf of his company, I took his words as an indication that there was a person who cared to do their job, and do it well. Without being himself and having a personal touch, it would have meant far less to me. It is also interesting to note that neither his avatar, Website, nor his profile background had the Sprint company image included.

Is this a dirty tactic meant to deceive? It shocks me how many people are skeptics and seek the most negative view on anybody in business. With social media, it is striking how many times you will hear Complainers and Antagonists telling you how to communicate with others and to never promote your business. To my notion, it simply requires that you be tactful. I believe that personal branding such as seen with @JGoldsborough and others shows an interesting realization on the part of large and small businesses. When people network with people, business becomes much more pleasant and effective.

Honesty is Best!

I should not even need a section on this, but let us look a bit deeper into honesty than just the difference in telling the truth or telling a lie.

Most of us have likely heard that lies are unsustainable. Parents teach their children lessons of honesty from a very young age, and you will find these lessons in everything from fairy tales to religious books. It seems that social media amplifies the importance of honesty. This honesty I mention can be as important as avoiding omission. Although omission is perhaps not an outright act of dishonesty, I encourage openness and honesty in your use of Twitter. This is not to say that you must tell everything there is to tell, but if you disagree with something, unless it is wildly controversial, it tends to be the case that those who disagree will respect you more for your honesty than a false agreement.

Do Not Fear Disagreement

We will not all agree on all things, but by having a position you will often find an even greater respect from others, including those who disagree. I wrote about "The Antagonist", but it is best if you remember that this is a relatively small and useless group when compared to those who will support you for being yourself and taking a stand. Just consider politics as an example.

Would you rather vote for somebody with a solid opinion, or one who tries to please everybody?

I have brought on attacks for something as simple as rebutting a woman who criticized a very nice inspirational video by saying it had too few women. I told her I thought it was petty of her to even notice. The video was inspirational, and after I sent it, it was spread widely by others in retweets. Yet, I received negativity from this woman. I did not appreciate it, and I am not afraid to speak out about it. This particular case exploded into rumors and ended up with statements that I was a jerk, a sexist bastard, and that I all but grabbed her by the hair and threatened her life. It was all so absurd to watch rumors start and to see others asking their friends to unfollow me that I had to laugh. After all, what kind of self-righteousness did they possess to justify telling me how to communicate? I simply responded to a negative comment by being just who I am. Incidentally, I lost an estimated nine followers in that outburst, but gained many times that amount. I also received astonishing praise from those who follow me and know me, and also from some I had never met.

Defending Oprah

Since I really want to drive this point home for you, I will tell you a couple other instances when disagreement came to be a benefit rather than hindrance. I inadvertently brought on many harsh words of disapproval when I gave my opinion on the topic of Oprah Winfrey caring for others in Africa. The argument that others made was that she should keep her money and her efforts in her home country of USA. My defense of her was that although they may consider it nice if Oprah did more in USA, that it was **her** business to run, and **her** hard work and success that made it possible for **her** to make those decisions. I also brought the point that humanity has no borders and that to be kind to others should be commended regardless of where the recipient is located. This one really brought out hate from some people. Maybe this will even make **you** dislike me, but I can assure you that if you ask me a question, you will be far more likely to hear what is on my mind than to hear me agree just so that you will like me.

In the instance of defending Oprah, I did not get a lot of praise from others, but what I gained was a greater respect for myself. It helped me to think more about kindness, and how placing a border on kind deeds is not kind at all.

Just one more instance of being open and honest is the story when Ashton Kutcher began following me on Twitter.

Ashton Kutcher Story

Much of America, and the world, heard about the popularity contest between Ashton Kutcher and CNN. Ashton proposed a challenge with the news giant to be the first to have one million followers on Twitter. He vowed to give $100,000 to buy mosquito nets for a fight against malaria if he won the challenge. My initial opinion was that his delivery came across as arrogant and could have done better with some coaching. That was my opinion, and I shared it with my friends on Twitter and on my blog (http://tfbtff.com/akcnn).

In my blog, I harshly criticized Ashton's watch party and his victory speech which was viewed by many in a live video Webcast on Ustream.tv. He talked about how he was changing media forever. I think perhaps the part I considered the most alarming was that he was claiming to be the little guy who went up against a media giant and won. After all, he is not really "the little guy" to me. I mean, he even had a billboard in my town, Topeka, Kansas. This guy has quite a large stage and a lot of dollars to be calling himself the little guy. *I* am the little guy. In some circles, I do not even qualify for this claim. So, I wrote about it and I saw a lot of agreement.

I also wrote of how he claimed that he was listening. His statement was *"There are a million people who need to be thanked, because I am not following me, I'm following you."* He was excited when he said this, and perhaps it was not the most thought out statement. However, I later realized that just because he is not following all of these people's every Twitter update, he is listening, selectively. In order to illustrate that he does give some attention, I offer the article as it was written in my blog as follows:

Kutcher vs. Murnahan Twitter Dance-Off

If you are not familiar with my blog, it may help to know that I wrote an article recently that criticized the highly publicized race between Ashton Kutcher (@aplusk) and CNN (@CNNBrk) to reach one million followers on Twitter. I look at this today and realize that nearly anybody could do that if they put the resource into the campaign that each of these giants did. There is even a billboard in my home town of Topeka, Kansas advertising to follow @aplusk on Twitter!

Ashton Followed My Twitter Feed

I have been asked a lot about the bar fight / dance-off talk on Twitter between Ashton Kutcher and myself. First, I want to say that there was no bar fight. Secondly, as of yet, we have not determined the venue for our dance off. However, Oprah's show (@oprah) seems to be the crowd favorite. So what is the real story? Since I wrote a recent post criticizing the race between Ashton Kutcher (@aplusk) and CNN (@CNNBrk) I suppose it deserves this follow up.

The Bar Fight

The story behind the bar fight is pretty simple. I sent a tweet as follows: *"20 Athletes You Would Want with You in a Bar Fight http://bit.ly/hxNg0"*. This tweet was subsequently retweeted by my friend, Phao Loo

(@phaoloo). Ashton Kutcher follows Phao Loo's tweets, and he liked it, so he retweeted it again. The bar fight story was really this simple.

The Dance-Off

Although there may be a dance-off, I have made Mr. Kutcher aware that I dance like a goon. However, I will be a willing participant if it should come to pass. Interestingly, the whole notion follows through nicely with some of what I believe his intended message was for CNN and Ted Turner. The message is that the little guy can still have a voice in social media, and can still make a difference in the world. The misguided part of that equation was that Ashton Kutcher is not the little guy. Unfortunately, Ashton did not have me on retainer at the time of his message, so a lot became blurred with many mixed signals and lack of solid personal branding.

Perhaps if Ashton Kutcher really wants to show the sincerity of his message, a better approach may be to come down to my level and visit with a guy who uses social media to build relationships and good will. Then perhaps he can win me over as an active part of his audience, and even his advocate. Beating all odds, he and I could work as a team to clear up his message and do some properly meaningful things together. As it is, I have been bombarded with questions of what really took place, and whether I think he is genuine. To all of the many questions, I must say that I simply do not know yet.

I have not made an attack on Ashton's integrity, but I have had reason to criticize his signal. In my previous blog post on the matter, I made observations that his message is flawed, and that I do not respect much of what I observed. Perhaps a lot of that has to do with Hollywood, publicists, contracts, and a whole lot of time spent being desired by masses. I have a strong immunity to the whole celebrity bit. I have spent enough time with "the desired ones", and also been the focus of some of the same. I do understand the awkward feeling of being envied, as certainly we all have for one reason or another. This transforms each of us differently on different levels. To me, it has made me work even harder to be real and to be humble. In Hollywood, this is a method that is often not embraced, or accepted properly when it is. After all, a little cockiness seems to make you cool. Sadly, the people it makes you cool with are the same people who are wearing a similar mask of blended pride and shame.

When it comes down to who the man is or what he intends, I do not know the answers about Ashton Kutcher. I will gladly welcome him to call me on my nickel, or join my Webcast as a co-host and come down to "the people's" level. I will gladly meet him for a dance-off and show just how horribly I dance.

The way it all went down was like a firestorm of tweets about a dance-off among my followers. Note that I follow well over 8,500 people's Twitter feed, whereas Ashton followed (at the time) 84. Not only do I follow this many, I do so very actively. As I write this, I have reached my 1,000 direct message daily limit and I am waiting to be able to send direct messages again.

Within my Twitter friends' rampant tweets, the speculation of a ghost tweeter was common, as were many RT, and questions of Ashton's intent (PR, damage control, etc). So that there is no speculation, the Tweets we have shared, and some of the related communications surrounding the bar fight / dance-off tweets are as follows:

@murnahan: Hey, Ashton Kutcher just RT me. http://bit.ly/Gt5Yf LOL!

@kimsherrell: OH SNAP ~>more drama. @aplusk just RTd @murnahan. but did he read mark's blog? http://bit.ly/4SBD8

@murnahan: ROFL! –> RT @KimSherrell: OH SNAP ~>more drama. @aplusk just RTd

@murnahan. but did he read mark's blog? http://bit.ly/4SBD8

@kimsherrell: WHOA @aplusk has challenged @murnahan to a bar fight?! ++++++++++++++++++++ +++++++++++++++++

@aplusk: @KimSherrell i don't have time to go to kansas. I just find the assumptions to be interesting.

@aplusk: @murnahan Kind of ironic?

@murnahan: @KimSherrell I guess he did see it. http://bit.ly/9iHpE

Right about this time is when I realized (from a friend's tweet) that @aplusk was following me. I was number 84 that he was following out of 1,234,083 following him. So the questions that came to mind was, whether he would still be following me tomorrow, and does this guy meet the criteria of those whom I prefer to follow?

@murnahan: @aplusk Are you an OK guy after all? I like seeing that you are listening. Dance off, huh? LOL I dance like a goon!

@aplusk: @murnahan lets dance brother. I think you paint interesting picture with your words.

@murnahan: @aplusk Thanks! Do you need any help at the TweetDeck? http://bit.ly/AKIra

@aplusk: @murnahan no I'm quite efficient i'm sure @oprah could use a tutorial though

@murnahan: Well then hook me up, brother. LOL –> RT @aplusk: @murnahan no I'm quite efficient i'm sure @oprah could use a tutorial though

@aplusk: @murnahan you wouldn't want to teach her she might ruin the platform by connecting with too many people

@aplusk: @murnahan or she might end up saving lives…

@murnahan: Well, @oprah, @aplusk said you may need a hand. http://bit.ly/2zog7w - REF: http://bit.ly/NHRbb

@kimsherrell: ashton kutcher now following mark murnahan. thank you + good night.

@kimsherrell: LOL… a low-budget internet musical: @aplusk versus @murnahan - http://bit.ly/dgyxa

@murnahan: Hey Ashton (@aplusk), I am following you back now, but don't go messing up my reputation.

There were many more tweets about this, but I must stop at some point. For more tweets relating to this, you may try a Twitter search and refine the search to find related information.

Will There Be a Dance-Off?

To answer the question of whether there will ever be a dance-off between us, only time will tell. I may never hear from Ashton again, or we may turn out to grow a mutual respect for each other. Who knows? To my notion, Ashton is a funny and talented guy. I simply do not believe that social media stunts last all that long, and

I am here for the long-term, just as I have been since the 90's with my Yahoo! chat clubs, where I met my wife.

What Happened Later?

As a follow up to this story, I should mention that I used Twitter to direct message Ashton multiple times. I asked for information about an effort he was a part of to help feed children at risk of hunger due to economic recession. I eagerly expressed my interest in knowing more about the cause. I stated that if it was vetted properly, that I was interested in promoting it or even giving a portion of the proceeds from the sales of this book to the cause.

I did not intend to promote the donation along with the book's release or marketing efforts. I would never want to be the guy who uses charity as a "Look at Me!" and a "Pat Me on the Back" venue. I have issues with those who blatantly try to help themselves more than the charity. You know the type I am talking about, right? This is something I once suspected Ashton of in my earliest criticisms. My thought was that if he really wanted to give, he should not need a million fans to look at him and pat him on the back. Just write the check the way so many others do, and feel it inside instead of seeking outside approval or taxation benefit. To me, that seemed like an arrogant trade off, with a selfish ratio of gain for giving. That was just my initial view. I have since tried to look closer and see the good in his efforts.

I let that criticism subside once I heard about the next charity he was promoting to help children. It was starting to look as if he may just keep helping, and this time, it did not appear to be self-seeking. I commend Ashton for his efforts. I hope that he will keep trying to help others.

I took my request for information about helping the cause for children pretty seriously, because I tend to be quite charitable when it comes to children. I am blessed with three children of my own, and I would never want them to suffer.

I eventually received a response, and I thanked him kindly when he thought to direct message me some time later with a link so that I could promote the cause. I did not receive communication on how I could help, other than to retweet a link that gave very little information. However, it was something, even if I did not get the whole picture of how I could help.

I reached out again to ask how it went, and whether he had any positive stories to share. Since Ashton has spoken of how social media has changed the world, I hoped that he would be willing to share this with people. I repeatedly offered the opportunity for him to provide positive influence, and to review my book before the launch, since he was to be included. All I heard was crickets chirping. I made a final plea when I direct messaged him as follows: *"I plan to send the book off to print within 24 hours. Last minute question: Are you listening? Did you get my requests?"* Still, just crickets.

He only follows 180 people on Twitter, and I sent this as a direct message. If he was following thousands of people, I could just assume that my messages were lost in the shuffle. This is not the case.

An Open Letter to Ashton Kutcher

I know that you are busy, Ashton. I respect that, and I know what busy is like. This is why I want to offer you this open-ended opportunity to show us all how the little guy can make a difference. I have a challenge to present to you. If you take the challenge, you will show that the little guy really can have a voice, that you really believe in social media, and make a charitable impact, all at once.

I challenge you to arrange a dance off between us on The Oprah Winfrey Show, and a talk about how social media and Twitter is really making a positive difference in our world. If you accept this challenge, I will donate a dollar of every sale of "Twitter for Business: Twitter for Friends" to a charity that will be decided by social media in a poll.

Your autograph on the line below indicates your acceptance of this challenge. Additionally, I would like to ask Oprah to autograph here as well, and offer this copy of the book to the biggest giver to the chosen charity.

Ashton Kutcher _____

Oprah Winfrey _____

For anybody who would like to see this happen, you may send a Twitter message including @aplusk, @oprah, and the hashtag #tfbtff.

Be an Individual Even When it is Risky

This is my take on being an individual. This is my brand, and you cannot have it, but hopefully this will inspire you to be yourself and to not be afraid to show what you stand for. This does not mean being an antagonist, and it is best to be respectful of others while expressing your opinions.

You do not have to take my word for it. Just consider what happened one day when I said a naughty word. I made a point by saying that I do not give a shit if somebody unfollows me for a petty reason. My message about caring more for relationships than pettiness was repeated in the form of retweets by many people. In fact, it was probably one of my most retweeted messages of the day, week, or even month. The response was overwhelming. It clearly gave people much enjoyment when I broke the ice and let them have a good laugh on me. I use this, again, as an example of never being afraid to be yourself. People will respect you far more than a facade of you.

The Twitter Retweet

The Twitter Retweet, although not a formal function of Twitter, is addressed in the official Twitter rules relating to providing attribution for messages that are not your own original content. If you repeat something that you received from another Twitter user, it is not only a polite way of passing the information along, attributing the content to the originating user is a Twitter rule. For more information on the topic, I invite you to see the URL as follows: http://tfbtff.com/snJ04

Being retweeted on Twitter is referenced on blogs, television, and magazines around the world, but is it all that important? I want to tell you a couple things about the coveted retweet, and why it matters.

What is a Twitter Retweet?

A retweet is when a Twitter user sends a message and another user extends it to their Twitter followers by sending the message including the statement "RT @username" or "via @userrname" showing the source of the information. If it is passed along multiple times, you may see multiple usernames referenced, which is appropriate, and accepted. It is common that the use of "RT" denotes a close resemblance to the original message, whereas "via" simply denotes that it was received by way of the user. It is also common to find that the message is either prefixed or appended with a user's own message. In this case, it is common to see a short message followed by "RT @username: Original message here." There is not a set right way to do this, but how I will often retweet a message may look something like this: "This is hilarious! RT @username: Check out this video http://_____.com/video". If I dramatically change the message, I may opt for using "via @username" and it may look more like this: "This hilarious video almost caused me to change my shorts! http://_____.com/video via @username"

Proper Reason for Retweet

A retweet is useful for sharing good information while crediting the originator. By retweeting, you are being courteous to the source. If you want to spread useful or interesting information, a retweet can be very helpful. There is nothing wrong with asking for a retweet, as long as the general purpose is for other's benefit.

The retweet expands the reach of a message to a level that you could never achieve with only your own resource. It is a form of syndication that can allow a message to reach an enormous audience. Because of the potential to reach many people, there are cases where the retweet can be misused and annoying. The people who provide little value and just want to sell you their goods would consider the retweet to be the Holy Grail ... the brass rings to reach for. I have seen cases where a user has used a retweet to inappropriately show perceived value by inserting "RT @username" into a message that was clearly not a retweet. I have also seen cases of retweeting something that a popular user sent, but replacing the link with something entirely different. If you find this happening, it is best to block the user and report the user by sending a direct message to "spam" along with the offending username.

Abusers of the retweet are often the same people who consider the number of people following their Twitter feed more important than focusing on the value they offer to their audience. These are (usually) not the most retweeted people. If selling is your only goal, you should really rethink the retweet.

Be Retweeted by Providing Value

If you hope to be retweeted, consider how you interact with people, and whether you are giving them what they want. People do not want to be "sold at". Imagine the difference in how you feel about going to a store to buy things compared to how you feel when the doorbell rings and there is somebody there to sell you something. This is very much how people feel about the selling invasion in their social networking space. I will address this more, later.

The methods and mentality of social networking has been written about in huge volumes. Some of the information is great, but much of it is junk. There are a lot of people writing about social networking just to sell to you something. Then, there are the ones who really "get it". The ones who do "get it" are the people who understand what I have said many times: *"If you see somebody as a sandwich, they can usually tell."* This is to say that if your intention is more about making money or being popular, and not about being helpful, most people will know.

The Value of Being Retweeted

I could tell you that the value of a retweet is that it will make you wildly successful, or that it will make products fly off your store shelves. I could probably even find some people who would believe it, but let's be realistic. That is not the way this works. Although it could conceivably happen that you have a product so desirable to the masses that the only reason you have not sold a *squillion* of them is that you never told Twitter. This is not normally going to be the case. What the retweet can do is to syndicate instances where you provided usefulness to the Twitter community. To me, I view it is a good measure of whether I was useful or interesting.

The Real Retweet Secret

I am going to put my neck out and say that the real secret to being retweeted comes in just a few parts. Here is a short list of the things I believe to be most important.

Forget about you: If you have the right mindset of being useful to people, rather than being selfish, it will show. People really like to talk about themselves. Let them, and don't hog all of that enjoyment for yourself. Listen to what people have to tell you, and you will probably find some great friends.

Forget the dollars: (see above) If people want to know what you do, they will find out.

Be friendly: Try to get to know your following, and allow them to know you. This is why we use the word "social". Address people by their first name. When you build a good reputation, and rapport, people give much closer attention to your updates.

Show personality: You are a person, right? Flaunt it, don't hide it. I often turn on my Webcast to have a chat with my Twitter friends. It has been great fun, and I doubt any of my friends would fault me for lack of personality, for better or worse.

Consider what people want: People want to connect with other people. They want to laugh. They want to find useful information. Things that promote what people want are very often retweeted.

Retweetable Length: In order for something to be retweeted, there must be room for it to be forwarded, along with an additional username or multiple usernames, if it is retweeted again. This means that you should start with a short message and not use the entire 140 character limit. If you want optimal retweets, you should keep it as short as possible and use a URL shortening service to keep Web addresses as short as possible. When a message becomes too long to retweet, it is best to abbreviate it, leaving the username(s) in place, and then add a link to the message as you received it. The link to the message is always available on Twitter.com in the timestamp of each message.

@murnahan Retweeted

I should explain that I did not make this all up. I have topped all of the popular leader boards as one of the most retweeted Twitter users, and remained there for weeks on end. Anybody can do it, and like anything else, it can also be cheated. All you have to do is visit Google Trends, Alexa Hot URLs, Digg.com, Stumble, or see what is trending on Twitter and write a quick blog about it. If you do this with high velocity, you will see a lot of people retweet your content. However, if you seek retweets, it is best to consider why it is important to you. I believe that a mindset of helping others and being useful is the best reason for seeking a retweet. On the other hand, if it is only to suit an ego, it is best to not waste your time.

Social Media: What's it really worth?

What is the value of social media marketing and social networking to a business? This seems to be a very misunderstood subject, and one which too many companies tend to relate to other marketing efforts or advertising. If you really do not see the enormous value of social media marketing today, then it is likely because you are either going about it wrong or you are not embracing it at all. It is time that we change that for you and help you to see what this social media thing is all about.

Social Networking ROI

I hesitate to use the term "social networking ROI", but since so many businesses want to know the value of networking using social media, it is a term I use for the company hoping to implement a social media plan. This could also apply to the growing number of individuals seeking new employment opportunities using social networking.

First, I want you to understand that much of what you learned about marketing and advertising goes out the window right now. This is likely not what you think it is … it is much more! If you try to measure a return on investment (ROI) of social networking, it is a lot like putting a ROI on each handshake or each "hello" as you walk through your grocery store. It is not about advertising, and the same metrics cannot be applied. The value of social networking flows down many streams, and it harnesses the value of good public relations, communications, marketing, friendships, and so much more. There is much that simply cannot be measured in a spreadsheet, and thus it will often take a degree of faith, especially if you hire a social media consultant.

Social Media is "Networking" and it is "Social"

The term "social networking" really says what it is, but I often find people who continually get it all wrong. Companies seeking to use social media for business purposes often get it wrong by trying to advertise their goods or services. This sets them up for an utter failure, and can often do more harm than good. Business people participating in social networks often find it hard to come out of their shell and actually be social. How do we fix this? Here are just a few tips about social networking that you should not overlook:

- Put your first name in your profile! People want to know YOU, and not a username or company name. Some people will still communicate with you, but is it really you?

- People you meet on social networks don't bite! Well, maybe a few do, but most will hug you rather than bite you.

- Get to know people. Spend time with them, just as you would with a friend anywhere else.

- Be yourself! I have said it a million times, but I mean it. We will not all get along, but most people will respect the individuality you show.

- Do not tell me about your business and then not remember my name! If you do this at your local

Chamber of Commerce function, or worse, at a cocktail party, you will look like a buffoon. You are not a buffoon are you?

- Remember that 100 people saying something nice about you to 100 other people holds far more value than a few business prospects. When you have friends, you will have people who want to be helpful to you. The value of a friendly referral can go a long way!

- Be respectful of others.

Social Networking Takes Dedication

If you cannot make a solid commitment to a social media plan, it is best to not start one at all. The people you will meet are real and you simply cannot make friends only to leave them behind. Just how seriously do I mean this? Let's examine a couple of my friends whom I met online and through these social networks I write about.

- **Peggy** (my wife)

- **Mike**, who is one of my closest friends, Web developer, and Kansas State Legislator.

- **Bianca**, whom I met online in 1998 and communicate all the time. She lives in Austria and used to be an au pair in the USA.

- **Eric**, who ate a 6 pound burrito, lives in Las Vegas, and sells bulletproof vests and armored cars.

- **Toni**, who blogs and chats with friends even while her husband, Royal, sleeps beside her in bed.

- **Melissa**, who nearly wets herself laughing at things I send her via Twitter.

- **Sail**, who lives in Honolulu, Hawaii and has not been to a beach in the last month.

- **Cody**, who lives in Calgary, Canada, likes to party, is a bodyguard to celebrities, and met a new love interest on my live Webcast (#Shabam, Cody!)

- **Misty**, who lives in Manila, Philippines and loves chatting and great design.

- **Robin**, who is a wonderful source of inspiration, and a true lover of mankind.

- **Linda**, who is a hilarious socialite from New York City, has the most amazing voice, and uses it for kindness.

This list can go on for thousands of people.

The people mentioned here and many more are among the very important relations I have met using social networking. I communicate with these people regularly online, on the telephone, and in person. They are all parts of a very important network of people whom I can count on to be friends and to have something nice to say about me, with or without prompting.

The Social Media Commitment

Just how important is a commitment to social networking? Like any wise investment of your time or money, you will benefit from social media commensurate with your efforts. If you have seen the news, it should be clear by now that other forms of communication such as television, radio, and newspapers are failing. If you have not heard this, it should be even clearer just how much their reach has degraded. Social media is picking up where the others have left off, and at the horizon is Twitter. With all of the huge changes in today's communication methods in mind, you should question how important it could be to your company to make a significant effort toward building a social network and social media plan before your competition does.

Can a Social Media Consultant Really Help?

This may seem like an alien question to many people. After all, if you are to be yourself, how can somebody be a better "**you**" than you, right? That is not what a social media consultant does.

When you venture into the unknown world of social networking, there are many pitfalls to avoid, and many useful tools that can help you. Having a professional on your side to guide you and to promote you can be a huge factor to your success in reaching the right audience, with the right message, and with the right approach.

The answer is YES! A social media consultant can be an immeasurable benefit to your social networking efforts, and should be chosen wisely.

Which Social Network Do I Prefer? Twitter of course!

My personal commitment to social networking and Internet marketing is greater than I would ever expect for one of my clients. However, I expect my clients to make a substantial commitment to their efforts in social networking. In order to explain how committed I am, I will simply say that I spend far more time cultivating relationships with my social networks than most people will ever spend in an office. This is not a 9:00am to 5:00pm process.

How Personal Is Social Media?

I have given examples of the personal nature of social media, including Twitter. If you still wonder just how personal it is, just keep reading and find out how people really receive the personal aspects of social media. This next story is my thanks for an overwhelming response to a very personal experience in my life. I believe it shows very clearly how much better we are received as people, as compared to our reception as a business entity. This is taken from a blog post I made shortly following the birth of my son, Jack Walden.

Jack Walden Teaches Social Media

Jack Walden Murnahan, Twitter's youngest user, sent a message to social media yesterday that we should all take heed of. The message that Jack Walden sent at only minutes old was not just as you would think, and it should prove to each of us the greatest of lessons in social media.

Jack Walden's message was delivered more directly to the heart of social media than even the most cleverly devised sales pitch or news story. It did not need to be spread around the world to every Internet user, and there was no attached agenda. It was simply the sharing of one family's very excited welcome to their little boy.

The Response to Baby Murnahan Tweeting

The responses have been overwhelming. As I made announcements of progress toward the birth of our son, the outpouring of excitement and love was more than I could possibly keep up with. I sat down today to write a personal thank you to each of the people offering their congratulations and excitement for us. After writing several hundred personal messages of thanks, I started realizing that I was actually losing ground. More messages were coming in by Twitter, Facebook, and email, faster than I could keep up. Beyond just that, I knew that I would soon reach the 1,000 direct message daily limit and 100 per hour limit for @ messages on Twitter. I have reached both of these limits before, even without having a new son.

What this incredible outpouring of support for and about Jack Walden teaches is that people really do care about people. Social media provides a means to reach into people's lives, get to know them, and share in their joys, defeats, likes, dislikes, and more. It allows for unique and often touching insights to people's lives, and for many of us, it provides great joy to feel a part of something bigger than ourselves. The power of friendship and caring is something that cannot be described in a single blog post, or in only a few lines. It is built over time, and built with trust.

Even if you skip the rest of this blog post, I hope that you will heed the message that a little boy named Jack Walden Murnahan has come to deliver about sharing in joys and pains of others, and the very deep-reaching power of communications with others that is so greatly enhanced through social media.

I will, however, since so many people have asked, share some of what lead up to Jack's birth, and give you a story of this piece of my life that has been very touching

to me. So read it if you like, and know that I have held
your many well wishes and congratulations very dear.

Jack Walden's Story of Social Media

A while back, I announced that my wife, Peggy
(@pegmu) would soon give birth to our new baby. Since
so many of our friends are spread around the world, the
Internet and social media is clearly the best way to share
our excitement and details with our friends and family. It
is a lot faster to make a baby announcement using
Twitter than to call each person to deliver the great
news. Plus, it is a great way to show the new baby
photos and video to the people who wish they could be
there but cannot.

One of the earlier announcements of our joy was our
Twitter Kids video. The video showed how our "human
resources department" (Peggy) was working on bringing
us more help to keep up with our work. If you have not
seen it, you may get a chuckle from it. This was a fun
video for all of us to make.

As the pregnancy progressed, I shared it with friends on
Twitter and Facebook. On April 1st, I shared that we
thought we would be welcoming our new baby that day.
This was not an April Fool's Day joke. Peggy was
having very regular contractions, and they were
increasingly strong. However, once she finally got too
tired to stand any longer, she went to bed and the
contractions subsided.

Several times since April 1st, we were pretty convinced
that it was time to meet the little one. It dragged on for a
long time. We were visiting our midwife weekly, and we
kept our fingers crossed that we would meet our baby
soon. On April 16th, we made yet another visit to our
midwife following a series of contractions that seemed
productive and getting closer together. Peggy was

having contractions as frequent as every two minutes. Norla, our midwife, promptly put Peggy on a monitor and checked dilation. She sent us home and said that she would not be at all surprised to see us back either that night or the next day. At this point, Jack was already a week late, and we were becoming concerned that we may end up in a hospital where they would require a cesarean section (surgical) delivery. This was a very frightening prospect for Peggy, and she hoped to avoid it.

That night, Peggy did as she had for days; she paced up and down our street, stayed on her feet, and hoped that gravity would help to enhance the labor. She finally wore out and had to go to bed. She was completely exhausted. She finally got some good rest, and I did my part to be sure the kids would not wake her too early. I wanted her to rest as much as possible because I was certain that she would have a very exhausting day ahead.

That morning, she walked with her mother around our neighborhood and went shopping, mostly for the walk. By about 1:00pm, Peggy said that some of the contractions felt stronger, but they were just short ones that went away pretty quick. I suggested that we call Norla, just to be safe. We described what was happening, and Norla said to come on in and she would take a look at her. As we left our home, it looked like things were getting more serious. Peggy had a couple of pretty strong contractions.

We arrived at the birth center at about 1:55pm and they checked her blood pressure, pulse, and the baby's pulse. All of the sudden, Peggy was hit with a really strong contraction … I mean really strong! Of course, I tweeted it with one hand as I held her hand and comforted her.

It became clear that it would not be very long before we met our son. Norla could tell that things were happening fast, so she told Peggy to go ahead and put on a gown and that we would not be leaving without a baby in our arms. I will save some of the graphic details, but Peggy went from being dilated 4cm to giving birth in under a half hour. She pushed three times and delivered our son directly into Daddy's waiting hands in less than five minutes.

Minutes after his birth, Jack was ready to send his first tweet. Jack's first official tweet was as follows:

Jack's 1st tweet: Hi Tweeps. I was born! #baby #twanic #whew (now press enter, Jack)

That message, and the ones leading up to it, caused a huge rush of support and congratulations that I have been shockingly unprepared, I did not expect so many people to listen or care enough to show their interest of compassion for our moments of joy. I feel honored by the warmth given to our family. As much as I want to respond individually to each person, I have provided this story to tell a bit about what happened for our family, and how deeply thankful I am to each person giving their support and love.

I owe a huge "thank you" to each of you. You really are the reasons that social media is great. You are the people who understand that the very best things in life are the people and relationships that you build. You are my social media rockstars!

More examples from my personal experience that prove the personal nature of social media are simple for me to tell. I met some of my closest friends, including my wife, as a result of social media. Here is a story taken from a blog post I wrote about just how personal social media can be.

Three Kids Prove Social Networking Works

Social networking has been analyzed, scrutinized, bastardized, and commercialized, but my family is proof that it works, and that it has worked for over a decade. If you will give me a moment, I will tell you why I am blogging about this today, and give inspirational credit to people I met and have built deep relationships with that have lasted for over a decade, and those whom I only recently met. I will start with today, and I will go back to the really good stuff when I met Peggy.

I sent a tweet on Twitter one day that read as follows:

"Social Networking Fact: I met my wife online in 2000 and we await the birth of our 3rd child in April. It works, I tell ya!"

I sent this tweet after an engaging blog conversation asking *"When will social media be 'ready'?"* I am never the guy to leave a quick one-liner on a blog because I am just not a link-spammy blogger. I would rather say nothing at all than to say *"Great article, it was really helpful."* As it should be, my comment was thoughtful, and it was engaged by the author, Caleb Gardner. Here is how it went:

Mark Aaron Murnahan: When somebody questions the ROI of social media, they have already missed the point. It is worse than the mentality of trying to measure the ROI on taking a client to a ball game or going to dinner. Building relationships should not be measured by dollars and cents. I have just been communicating with a friend whom I met and have built a strong friendship since 1998. I have never asked her for business or for referrals, but you can bet that if she knows somebody who needs my services, I will get the call. Further, I did a lot of dating online years before it was widely accepted and met my wife online in 2000. We are now expecting our third child in April 2009. Social networking has been ready for years, but people being ready for it is another story. Social networking used to happen in ballrooms and the corner restaurant. The only thing that has changed is the venue.

Caleb Gardner: @Mark I love the personal story about your wife. Way to make an emotional (literally) appeal for social networking.

It's an interesting thought that social networking has been around for years. You're right - it literally has in that we've always built relationships with those around us. I think what is happening is that the Web is making us more cognizant of the relationships we build, because we're able to build them with people that we never would have been in contact with before.

Hmm... have to give that some more thought. Sounds like an interesting post on its own...

Mark Aaron Murnahan: @caleb Since my comment, I was on the phone with a good friend I know from my "other job" racing cars. His very financially successful company has a churn issue because of a hugely competitive market with tiny margins. I used myself as an example with him. I explained that he would not hear my message as clearly if he did not know me, my wife, my children, and my integrity. He has been in my garage working on racecars with me at 3:00am before a big event, and we drive around corners at 100+++MPH together, for the sake of Pete. He knows that I have a lot more at stake than a sales pitch. We have a relationship. I have tried to reach his executive staff to understand that without relationships, all we have is a sales pitch, and that people do not buy the price tag but rather what is attached. He gets the message, and he is really excited to work together, as am I, but he is getting a lot of pushback on implementation from his fellow execs. They have a corporate stuffiness that does not even match their written message and their goals. The bottom line is that if we miss the relationships, we work much harder and achieve less. You built on our relationship by engaging me with your reply. This is how stuff really works. My next blog article about it is forthcoming.

How I Met My Wife: Retweeted

Remember that tweet I sent? It was retweeted and replied to, which is always an honor, because it means I said something that people actually heard and thought enough about to tell others, and to ask a question. A question was posed by @askorkin as follows:

"really! you met your wife online, i am intrigued how? if you don't mind my asking :)"

I replied: *"Since you asked how I met my wife online, I am inspired to blog about it. It is my best proof that social networking really works."*

My Social Networking in the 1990's

Once upon a time, I had a friend and business partner who did not really understand the reach of the Internet. He was a physician and I was a marketing guy. We were working on a project targeted toward pharmaceutical companies that were spending tons of money to bring doctors to luxury resorts in Miami, Palm Springs, Orlando, Phoenix, and elsewhere, to learn about their new drug. At the time, thanks to government subsidized travel and tourism in Central Europe; we found that it was actually less costly to bring American participants to a conference in Budapest, Hungary than to Miami. This became the target of our new conferencing company.

Jeff posed a lot of questions as to how the Internet worked into our business model. He did not really "get it". I explained that I had developed a network of friends in the region, and globally. Even back then, my European social network of friends included Bianca, who was an au pair from Austria working in USA, with whom I have communicated even in the last 24 hours.

We made connections with hotel managers, tourist attractions, and one of our favorites, Varsaci Karoly ("Karchi"). Karchi worked for the E.C. as a Euro Qualifier, and we were fast friends. We got to know him online, but he soon showed us many incredible times in Budapest. We had a lot of fun at the courtesy of the Hungarian government. After all, it was ideal for them to attract American dollars back then.

It was really sinking in for Jeff, by this point, that this Internet thing could be useful. However, he still questioned it as a marketing tool. It is funny how most people think of it as a marketing tool first and a networking tool second. We (he) flipped that around. Jeff knew that I was pretty "Internet savvy", but I needed to give him a clear example. (This gets to how I met my wife.)

Social Networking Study: "Mark the Single Guy"

In 1999, I was single, 27 years old, and retired long enough to realize that retirement was not a good idea. My marketing business had been pretty good to me, but my personal life was lackluster from years of focusing on my work. Jeff's challenge to use the Internet to show localized results led me to kill two birds with one stone. I wanted a woman to hold and he wanted to understand the Internet. Again, it is funny how he thought localization was the challenge back then, but now globalization is the challenge.

I set out to prove that there were enough people right there in our town of Topeka, Kansas, USA to show the

Internet market reach. Of course, back then, his concern was that the audience may be too slim. Wow, I showed him. I used a XOOM.com account (back then xoom.com was a free host) to create a significantly detailed biography of "Mark the Single Guy" and I used a ".cjb.net" account to shorten the URL. I included everything I liked, didn't like, and I even had a special section about my baggage. The "Mark's Baggage" section was complete with an affiliate link to ebags.com. I promoted my heart everywhere I could, and I used the equivalent of today's "retweeting" by asking my online social network to pass along the bio and help me find the woman who would become the love of my life. At the time, Yahoo chat clubs were a big hit, and helped me to meet many friends.

It was not too long before I was receiving 300-400 email messages per day from ladies within a 100 mile radius who wanted to meet me. I met a lot of ladies, and I had a lot of fun … yes, a lot! I met a few neurotic ladies like Sara, Nancy, and DeeAnne, and I broke a few hearts. I am still sorry for making one of "the Stacys" cry. She was a sweet girl, and I really liked her family. Of course, Sara, Nancy, and DeeAnne were the ones I really wanted, but thanks to them, I was single when I met my darling wife, Peggy.

After the neurotic gals had nearly broken my will, I was pretty careful when it came to being close to Peggy. I think I was in love with her before I ever touched her hand or smelled her hair. Peggy was clearly very special, and I would do my best to keep from hurting her with my baggage (and WOW, I had baggage). Peggy does not like to admit this, but she admitted back then to crying as she read the deepest parts of me, the man she really wanted but was afraid of.

Shedding My Skin for Peggy

Sharing the real me was like pulling a scab off my entire body and letting air hit my sensitive inside. In my biography, I had shed my skin and stood emotionally naked for the world to see and inspect. It is lucky for both of us that I was real. I showed my sweat, my tears, my fears, my body odor, and the things that made me a real person.

Is This Fact or Fiction?

Some will question if this is all real, just as Peggy did back then. The accounts you have read here are only a small part of the full story, but it is all real, and it is all me. Perhaps as you get to know me, I will tell you the really deep and hard stuff that I once shared more freely online.

It cannot be all wrong to share who you are. After all, that is how I met the love of my life and the mother of our six year old son, four year old daughter, and one month old son.

If you would like to know me better, just tell me so, and we will make that happen. I know the value of social networking, and I treasure the many relationships I have built.

Advertising Using Twitter

Advertising is an extremely sensitive topic for many users of social media. There is an adverse hypersensitivity to advertising in social media that often has a backlash that is wise to be avoided. However, after putting a lot of thought into this, I have realized that it is O.K. to advertise using Twitter. Yes, that is right, I just gave my go-ahead to advertise whatever it is that you sell! Of course, you did not need my approval, but since you are reading my book, I hope that it gives you a bit more confidence.

For the people reading this who know me, go ahead and pick your jaw up off the floor and keep reading. I have been using social media for a good many years, and one of the biggest complaints I hear is about people advertising. I have been one of those people complaining, but now I realize that the one thing worse than the advertising is all of the people who complain about it.

Self-Promotion: The Dirtiest Word Ever

In the lexicon of social media, I would give my vote that the term *self-promotion* is the ultimate insult, depicting the vilest scum of the Internet. There are many reasons that people will give to complain about promoting oneself, but for each of the complaints, I can find just as many or more reasons that complaining about it is futile, if not just as asinine.

Let us examine for a moment which is worse, and which you hear more complaining about. In my experience, I have found that I witness more people complaining about self-promotion than I actually find self-promotion. It seems that for every one act of self-promotion, there are several people complaining about it. So how does this make the matter any better? When you complain about it, you are not likely to impart your ultimate wisdom upon them and change their business practice. Sure, you can imagine that when enough people complain, that it will shift the tide, but that is unlikely. Consider how much the huge voice against email spam has helped reduce the junk in your inbox. It is still the majority of all email, whether you stomp your feet and get mad about it or not. The difference with social media is not that you have a greater voice to change it, but simply that you have a greater voice to show frustration.

A likely scenario of complaining about self-promotion is that you will bring about negativity that others will join in with and perhaps make you feel stronger and more consolidated, but is it productive? My experience is that it is not only a waste of energy, but also promotes negativity. In a worse scenario, you may bring even more attention to the promotion, and for the person using this as a tactic, they often believe that any attention is good attention. Following this logic, complaining only fans the flame, rather than blow it out.

On the topic of self-promotion, you would perhaps do better to refer them to my book than complain. Now how does that strike you? That was an appropriate self-promotion. If I have value to share, do you really mind that I make you aware of it?

How Will They Know What I Offer?

This is where my earlier statement regarding advertising being O.K. comes true. I really do believe that it is fine to tell people what you do and what you offer. The most effective way I find to do this is to make many friends and let them talk about me when they find it appropriate. That is called networking, and a warm referral from a friend is much more valuable than an advertisement. Just imagine how many more people I can reach when I consider the importance of selling my services to referrals from friends compared with selling to my friends.

Of course I want my friends to know what I do, so I will tweet useful information and direct people to my blog. I write a good blog, and I do it to share good information that people can use. I am providing a value to others. I invite questions, and I provide my responses. There is a value proposition rather than a waste of my readers' time. I have tweeted that I am accepting new clients, and I have requested others to tell somebody about me if they know people who can benefit from what I offer. When somebody respectfully asks me to pass along their name, I do not find that to be insulting or rude, nor do most other people.

If people want to hire my services, I welcome it, but I realize that there is a far greater volume of business out there for me in the networks of friends than the comparatively small number of people receiving my tweets. In my experience, this is the greatest benefit you can receive from social media marketing.

If you want to simply advertise without being inconvenienced by having friends, be aware that it will not provide the greatest result. Some people will advertise by simply tweeting about their latest sale. You can generally tell this when you click the link to follow them. If you do not like it, do not follow them. On the other hand, if you are planning to purchase a car and there is an automobile dealership that tweets a description of cars they have for sale and their latest sale price, you may find great benefit in following them. I think it is fine for the coffee shop or restaurant to tweet their daily specials or similar uses. I simply understand it for the value it represents, and I do not hold a negative view of them.

Each user has their unique method and purpose. I have created Twitter accounts simply to make announcements about my racing Webcast or storm chasing Webcast, and not for two-way communication. Any people following those accounts were aware of that when they followed.

Time Magazine on Starbucks

As I wrote this chapter, I observed a new criticism on advertising using Twitter. In this case it is an article speculating on the potential backlash of Starbucks Coffee using Twitter to promote their product. The article is titled *"Starbucks Brews a Plan to Twitter for Dollars"* and is worth a read. Find the article at the URL as follows: http://tfbtff.com/8tXad

I see things every day pointing out somebody's misuse of social media. I wrote about this in the types of Twitter users, and I call them "The Complainers". In my opinion, Starbucks should use their social media the way they see fit. Freedom to use our words is arguably the greatest right afforded by the United States Constitution and an important factor in the popularity of social media. The Complainers have their rights, too.

How Twitter Improves Website Exposure

Nearly anybody with a blog or other Website that is worth reading has heard of Twitter by now. The bloggers who use Twitter efficiently may already have realized much of what I will share here, but from what I have found, many bloggers have not. To say the least, I am shocked just what a small percentage of blog owners and authors are actually using Twitter, and even more shocked by those who are not embracing the synergy the two can produce when used properly. Note that while I say "blog", this is completely interchangeable with "Website", so don't be confused.

Twitter-Improved Traffic: A Simple Example

As a simple example of how Twitter can improve your blog readership, I will use my new blog, aWebGuy.com. While looking at my statistics to find how many people arrived at my fledgling blog via Twitter, I found that an estimate of 15,000 unique readers arrived by way of Twitter in April 2009. I say estimated because it is challenging to provide an exact number due to the many readers arriving by way of Twitter clients not providing an accurate source. This was more than the number of subscribers to my Twitter feed at the time. So how did that happen? In short, people read it, talked about it, and retweeted it.

Blog Traffic Numbers: The Real Scoop

I find that too many people find it convenient to try and lie about the real traffic of their blog. Let's face it, this stuff is tracked. Some people may say that traffic estimates generated as a result of Twitter is bloated or that it is not so great, but just a simple look at Compete.com or Alexa.com will reveal a lot of what I am telling you. Of course, these are usually a bit behind and do not reflect an exact accounting, but they are usually reasonably close. I share this with you because I want to provide a real example for this purpose.

When considering the 15,000 unique readers arriving from Twitter in April, let's weigh in the facts that the blog was launched in December 2008, and my Twitter account only had 78 followers on February 7th of 2009. It is a targeted blog about social media marketing and search engine optimization. I am not seeking everybody to read my blog, but rather the **right** people to read my blog. Be sure to also look at your own blog / Website and keep a close eye on the numbers and percentage of increase coming from Twitter. It is pretty eye-opening what Twitter can do when used properly.

The traffic and level of engagement prompted me to question how Twitter has changed blogging. Here are just a few things that I found.

Twitter-Improved Reader Engagement

I will show examples of blog reader engagement separated into three parts, but this remains only one of the three areas of benefit derived from Twitter discussed here. There are many ways to determine reader engagement with a Website. Some ways blog authors have traditionally found valuable to measure the reader's engagement are as follows:

Time on Page: The average time the user spends on a page is a good measure of whether they are actually reading what you have to say. This is clearly subject to the type and length of content you provide, but in any case, readers who are not interested will not stick around very long. A minute is a very long time for many internet readers. What I have found in the measure of time on page is that readers initially engaged by Twitter will spend more time reading my blog, totaling about three minutes per page view. This is a significant increase over users arriving from other sources, and is longer than any other source.

Page Views Per Reader: The number of pages each user visits is a strong reflection of the user's interest in your industry, beyond the single topic of the initial page they viewed. I have noticed an improved page views per reader coming from Twitter, up .5 page views per reader compared to other sources, which is a significant sign of reaching the right audience.

Blog Comment Volume and Quality: An important measure for the blog author is in how many comments, and the quality of comments the blog post receives. When writing something relating to Twitter, I have witnessed great results for blog comments, on my blog and others. One of my recent blog posts relating to Twitter usernames has received upward of 150 approved comments. I attribute much of this to the fact that many readers already have some knowledge of the author, and are already a part of a conversation. A blog is one way that they find out more information and continue the conversation. I think many bloggers would agree with this finding.

Assessing the reader engagement of Twitter users, both on a blog and on your Twitter feed, can also be measured by the comments received in reply to the posting of the blog link to Twitter. I find that some people will respond to the title of the tweet. It seems that every day I see somebody respond to the text of a tweet in a way that I know without question they have not read the blog post linked to the tweet. A great example of this was when I tweeted a blog post titled "Will Oprah (@oprah) Ruin Twitter?" and I received a lot of comments in defense of Oprah Winfrey's use of Twitter. That was kind of silly, because the blog discussed the changes that may come from the inevitable increase in traffic and how a large influx of new users may change how we use Twitter. This absurdity should always be considered a measurement of engagement of your Twitter following and not of your blog. What it also points out is that Twitter users who do read your blog are likely truly interested in what you have written. If your Twitter account is managed properly and you spend time to get to know your followers and let them know you, blind comments should largely only happen with your newest followers.

Twitter-Improved Search Engine Optimization

Many search engine optimizers (SEO) will overlook the value of Twitter for improving search engine penetration. If they miss this part, they are making a big mistake. A reason many SEO will dismiss this value is that Twitter uses the "nofollow" attribute in outbound links, thus, no increased Google PageRank. Make no mistake; Twitter can greatly enhance your visibility in search engine results. This can come from many outside factors related to Twitter, as well as Twitter itself. I will just name a couple, but here are some ways SEO is enhanced by using Twitter.

I should point out that Twitter's Search is a Search Engine. As more people use Twitter search to find information, using Twitter will help many people to find your information. Aside from just Twitter searches, the likelihood of particular tweets being listed in other search engines referencing a Twitter tweet or one of the many Twitter-related applications along with your link are improved. Be sure to realize that each person who reads your blog also comes with a voice to further spread your blog in many other ways. Thus, each reader who finds your blog in any Twitter-related way has the potential to further propagate your message in search engines as well. It all adds up to make a significant end-result.

Twitter-Improved Call to Action

Along with the added benefits of brand recognition and brand loyalty, comes the greatest benefit of all … an improved call to action. This means that the message you distributed has gone beyond just readership, and the reader has heard and responded to your call to action. In my case, that literally means that they have made a call to reach me and discuss improving their market reach. For you, it may be that they enter their order for your product, apply for a job, donate to a cause, or many other possibilities.

How I measure a greater call to action from Twitter: It has become common that I speak on the telephone or on Skype to a minimum of five different inbound callers per day as a result of Twitter users who also read my blog. In addition to inbound callers, I also call at least five people I meet on Twitter to simply make an introduction and to get to know them better, so this certainly works both ways.

I make it my practice to reach people beyond the singular communication tool of Twitter, and expand my communications to other tools. This means that not only has Twitter greatly impacted my blog readership; it also goes far beyond blogging to reach people I would likely have never met otherwise. It also allows me to do this in a highly targeted manner. As mentioned earlier, this could be geographically, by industry, niche, or etcetera.

Hiring a Social Media Marketing Consultant

If you want to take a well-conceived approach to social media marketing, it may benefit you to hire a consultant to help you. A consultant can create a social media strategy, train staff members, help with additional staffing, or manage a campaign entirely. If you are unsure, it is best to ask around. There are good and bad consultants out there, just as with any field where a lot of money is at stake. I welcome you to contact me for services or to refer you to others whom I know and trust. Something you should know about the good social media consultants is that they are generally in it for the right reasons, and will do their best to help you find a good fit.

If they have what you need, and take the time to prove it, the job is then up to you to recognize it and be ready to strike a deal. A quality consultant is your best friend, and you will be lucky to have him or her to work with.

The Easiest Field to Hate Since Attorneys

There are a many people trying to get their hands on your hard-earned money and calling themselves social media experts. If you want to know their effectiveness, ask them for their experience, and ask them for proof. This will shut many of them up right away. The reason I say that it is the easiest field to hate since attorneys is pretty clear. There are more bad ones than good ones, like any field overcome by greed. There are also many inexperienced marketers trying to "fake it until they make it." Find out what they know before you sign a check. If you get a good one, keep them close, because the good ones have a lot of benefit to offer to your business.

How the Big Dogs Get Paid

Now I am going to tell you how the "big dogs" in social media marketing get paid. I hear it all of the time, and many of my peers say they hear it, too … "so how do you make money with social media?"

Please note that this relates to social media consultants who earn their living by helping clients with marketing their products or services better with social media. There are many branches to the field of social media, such as bloggers who earn money as writers and blog owners. There are providers, such as Twitter, Digg, Sphinn, Linked In, Reddit, Facebook, and etcetera. There are social media marketers who saw some success in the industry and decided to write a book about it and sell it to make money, and similarly those who speak about their success in public forums. These are various methods, and this is not about them.

What I am writing about here are the social media marketing people you encounter and wonder how they earn a living. The "big dogs" are the ones who seem to have a lot of connections, including many thousands of followers on Twitter and elsewhere, and who seem to always be active with socializing in social media.

Big Dogs Love People!

The big dogs of social media marketing really love people. The most successful social media marketers are the ones who would give you the shirt off their back and ask if you need their shoes too. The biggest dogs with the best pedigree in social media marketing, and with the biggest social media respect, are the same people who can laugh with you when you tell them a funny story or counsel you when you have a bad day. They may not reach each and every person, but they sure try, and they are sincere. They care about you, and you are not a part of some under bellied marketing plan.

Big Dogs Get Paid to Have Fun

It is really amazing, but yes, social media marketing big dogs get paid to know people, make friends, and have fun. So, you may think that is crazy, right? Let's examine this. The big dog of social media marketing consulting does not look at you as a meal ticket or a box lunch. They want to know about you. They want to hear from you and have a feel for who people are. The big dog of social media has a genuine enjoyment of being your friend.

The Big Dogs' Agenda

So the business side of the big dog comes out, and I will tell you their agenda. The agenda of the social media marketing big dog is to know people. The big dog uses an understanding of people (yes, including you) to know what people want. Once the big dog knows anything and everything about what people want, they use this information to help their client (the people paying them) to be a better company and to best express themselves to the appropriate audience, and to do so more abundantly, providing a greater return on their investment. By the time the social media marketing big dog is ready to bring something to market, he or she has polished that offering to be positioned at the top of the given industry.

Big Dogs Run in Packs

A pack hound mentality is not really as ugly as it may sound. Here is how it happens: Once the big dog has done their homework and knows the perfect way to reach the people who will best benefit from their clients' product or service, they search their brain and their contacts to seek assistance. The big dog will likely make telephone calls to other big dogs to ask for advice, and for references of who they know that can help. This may be to find an editor at a popular industry-related news agency, blogging site, or other periodical. It may also be a series of email and social media messages such as Twitter, Digg, Sphinn, LinkedIn, Reddit, Facebook, and etcetera. If the big dog has done their job well, and has improved their clients' message to one that is appealing to a massive audience, they will put their reputation on the line and ask other big dog friends to pass along the client's revised and shiny new image directly to their networks.

Social Media is Easy

So here you have it. Social media big dogs get paid for a lot of fun, doing what they truly love to do. It is really easy in some ways. You may wonder why they get paid for it at all. The truth is that although it is a lot of fun, the social media big dog also uses a lot of social equity, analytical marketing experience, media insight, and much of their time in making their clients massively successful. The social media big dog makes many efforts to help their clients understand where their offering should be positioned in the marketplace. They often train and coach key client personnel in proper public relations, keep watch over the client's reputation, open doors the client never realized existed, and much more.

What Can They Do That a Client Cannot?

The social media big dog often encourages individuals and companies to do everything they can to engage in social media. They try to give their best advice, and hold very few secrets. However, even with all of their coaching and training, and even their list of contacts, a well polished message will often still fall on deaf ears. The relationships and the experience of any big dog will vary, but you can be assured that a real big dog has put in the time and effort to build an invaluable network. If it is truly a big dog, you have one that you are wise to not let walk away. That network sits silently behind them when they are at your bargaining table, and they can prove it to you.

A company can do this on their own, but it is often the equivalent of a father going to school for dentistry in order to take care of his own family's teeth. When it calls for a professional, it is best to hire a big dog.

Hiring a Social Media Big Dog

It seems that most social media marketing big dogs are pretty busy. They usually have a lot of work to do, even when it does not relate to a specific client. This is because their job has a lot to do with building and maintaining relationships, and a constant study of the world around them. It also seems that many are not really into selling their service, but rather educating. After all, their job is far more about making a product or service so attractive that selling is not necessary. This can be misleading to both the big dog and the potential client. When the social media big dog sniffs you, they want a feel for the culture of your product or service. If it smells bad, they will likely walk away without any further interest. This is because from their vantage point, they recognize the greater value of their reputation above that of your money.

As you read this book, it is likely that it reached you because a dog broke his chain and helped deliver this message to you. The author runs with a lot of amazing hounds.

This is my job as a social media consultant. I enjoy it very much, and I love to help others. If you know somebody seeking to make the most of their business, I am only a tweet away.

Twitter Cheat Sheet For Your Pocket!

Now that you read the book (you read the book, right?) you
probably forgot some of the basic information that I have shared
with you. Don't sweat it. The next couple pages are intended to
be a handy cheat sheet that you can either bookmark or tear out
and shove in your pocket or keep at your desk for quick
reference.

Basic Twitter Functions

Direct Message format: D username message

Example: D murnahan How are you today?

Note: There is a space between "D" and "username", and no "@" symbol. The recipient must be following your Twitter feed before you may send them a direct message.

Public Message Format: Simply enter your message and send. If you wish to include a specific recipient or recipients, enter their usernames as @username1, @username2, etcetera.

Example: @murnahan @tfbtff How are you today?

Note: There is no space between the "@" symbol and the username. The placement of the "@username" does not matter. If a message contains "@username", the user(s) will receive it as a mention.

Search Functions

Search for this:	Returns results for:
this that	both "this" and "that"
"this that"	exact phrase "this that"
this OR that	either "this" or "that"
#this	the hashtag "this"
from:username	tweets sent by "username"
to:username	tweets sent to "username
@username	tweets mentioning "@username"
near:London	tweets occurring near London
near:Paris within:10mi	tweets within 10 miles of Paris
this since:2009-07-25	"this" since July 25th
this until:2009-07-25	"this" prior to July 25th
this -that	"this" but not "that"
this ?	"this" and a question

this filter:links   "this" containing a hyperlink
this source:seesmic   "this" sent through Seesmic

Useful Web Address List

Below is a list of useful Web addresses used in **"Twitter for Business: Twitter for Friends"**. I have shortened them for your convenience. In any instance where the address changes or I discover a service that I believe will be more useful, I will update the shortcuts listed here with a comparable service or a blog post with an update to assist you.

Twitter for Business: Twitter for Friends   http://tfbtff.com
TFB:TFF Updates   http://tfbtff.com/subscribe
TFB:TFF Webcast   http://tfbtff.com/live
Twitter Help   http://tfbtff.com/help
Twitter Search   http://tfbtff.com/find
Twitter Advanced Search   http://tfbtff.com/advanced
Hashtags   http://tfbtff.com/hashtag
Trends: Retweet Radar   http://tfbtff.com/radar
Trends: Twemes   http://tfbtff.com/twemes
Follower Management   http://tfbtff.com/follow
Analyzer   http://tfbtff.com/analyze
Polls   http://tfbtff.com/polls
Account Backup   http://tfbtff.com/backup
Seesmic   http://tfbtff.com/seesmic
TweetDeck   http://tfbtff.com/tdeck
Mobile   http://tfbtff.com/mobile
Posterous   http://tfbtff.com/Posterous
FriendFeed   http://tfbtff.com/ff
Localization   http://tfbtff.com/local
Twitter Demographics   http://tfbtff.com/demographics
Success Stories   http://tfbtff.com/success

Twitter Success Stories

In the spirit of social media, I asked Twitter users to tell their story of how Twitter has made their lives better. When I requested stories, I thought it would be great to include the stories in this book to show you how others value Twitter. After some deliberation, I realized that if I placed a deadline on receiving submissions, I would miss some really great stories.

I want to share these stories with you, but even better, I want you to become a part of the story. Please join me on my blog at http://awebguy.com and on the blog dedicated to this book at http://tfbtff.com and share your story of success.

I want to give thanks for the early-responders. You may read their stories at the URL as follows: http://tfbtff.com/3ZI02u

To Jenna, Robin, Jan, Janet, Joseph, cheth, GlemiGlider, Logomotivemike, misslindadee, Dave, and others with a story to tell, I want to thank you for being the inspiration for others to join us for the biggest conversation ever that we call Twitter.

The Conversation Starts Here

When I set out to create a book about Twitter, my greatest goal was to provide a work that would be sustainably useful. I decided that I would not publish it unless I believed that it could be helpful to the public, including those hearing about Twitter for the first time, and also existing Twitter users.

I wrote this book as a living document. This book is just the tip of an iceberg about the use of Twitter and the revolution of social media. There is much that we have to learn from each other, and so the conversation has just begun. The next step is to reach out with the tools and ideas that I have shared here, and be the best *"you"* that you can be. As you grow your social media relationships, I hope that you will find this information to be helpful.

Those of us dedicated to building and maintaining the community of Twitter must be vigilant. As with any community, you will find those who will participate constructively, those who provide little benefit, and those who damage the community. We each decide the role we will take. I am asking you to be a participant.

If you like what you found here, please help me to spread the word about "Twitter for Business: Twitter for Friends" and

subscribe to receive updates by RSS or email at http://tfbtff.com.
I also invite you to join in my frequent live Webcasts and Skype
conference calls at http://tfbtff.com/live or
http://murnahan.com/live where you may participate and get to
know some great people.

The Beginning

Notes:

Tips: Is there somebody you would like to find on Twitter? Who do you know on Twitter that you have not communicated with in a while? Who should be using Twitter but is not? How will you use the information you found in this book? Is there an advanced search that you should monitor? Do you have questions to pose during the next "Twitter for Business: Twitter for Friends" Webcast and conference call? Are you following @murnahan?

_____

_____

_____

_____

_____

_____

_____

_____

_____

_____

_____

_____

_____

_____

_____

_____

_____

_____

_____

_____

Notes:

Notes:

www.ingramcontent.com/pod-product-compliance
Lightning Source LLC
Chambersburg PA
CBHW060905280326
41934CB00007B/1200